Digital Products
Living Data is the Future

Edited by
Prof. Dr. Werner Dankwort
Universität Kaiserslautern
Prof. Dr. Josef Hoschek
Technische Universität Darmstadt

based on
the International 3rd Workshop
in Current CAx-Problems
held at Maria Rosenberg (Kaiserslautern),
Germany on September 20th – 23rd, 1999

B. G. Teubner Stuttgart · Leipzig · Wiesbaden 2000

Die Deutsche Bibliothek – CIP-Einheitsaufnahme
Ein Titeldatensatz für diese Publikation ist bei
Der Deutschen Bibliothek erhältlich.

ISBN-13: 978-3-519-02645-7   e-ISBN-13: 978-3-322-84821-5
DOI: 10.1007/978-3-322-84821-5

© B.G. Teubner GmbH, Stuttgart · Leipzig · Wiesbaden 2000

Der Verlag Teubner ist ein Unternehmen der Fachverlagsgruppe BertelsmannSpringer.

ISBN 3-519-02645-7

Preface

In addition to the classical needs, competition on the global market requires from industry product innovations: quality, time to market, reduction of costs (Q,T,C). The modern process networks of product development and manufacturing passing the borders of countries and including several companies could not work without an extensive use of information technology. This is going far beyond the former idea of Computer Aided Design. Thus the 3rd Workshop on Current CAx-Problems did not focus on functionalities or methods aiding design like in the first two workshops but on "Digital Products - Living Data is the Future": problems of the virtual simulation of the entire industrial process, starting with the development of a product and covering the complete life cycle.

The workshop aimed at bringing together the three groups: industry (mainly automotive manufacturers), system suppliers, and fundamental research. During the workshop, communication between these three groups had to be intensified, and especially also among competing companies of the same branch to pave the way for concerted actions, which are essential for all in the future.

Therefore, presentations given by various participants regarded the following innovative topics:

- Process chain and target driven design
- Digital Product and DMU: State-of-the-art and requirements
- Dynamical Product Data Model for the whole life cycle
- Interoperability and future trends in CAx system architecture
- CAx tendencies and long-term strategies in industrial application

There were representatives from automotive companies and their suppliers: BMW, Bosch, DaimlerChrysler, Ford, GM/OPEL, Hella, Nissan, Porsche, Toyota, Volkswagen, and system suppliers: CAD-FEM, CAD-Technologie, CAxOPEN, CoCreate, Conweb, Dassault Systèmes, EDS, Holometric Technologies, IBM, Imageware, Nihon Unisys, Parametric Technology Corp., ProSTEP, SDRC, Software Factory, Unigraphics Solutions as well as from research institutes and Universities: RWTH Aachen, TU Darmstadt, University of Erlangen, FHG-IGD Darmstadt, FHG-IPA Stuttgart, IMA-CNR Genoa, University of Kaiserslautern, Research Centre Karlsruhe, University of Karlsruhe, University of Parma, Polytechnical University of Catalonia.

The participants were of different nationalities. They came from France, Germany, Italy, Japan, Spain, Switzerland and the USA, and the very familiar atmosphere in the Convent Maria Rosenberg fostered the fruitful working discussions as well as the possibilities to become personally acquainted with one another. In a final panel discussion the contents of the workshop were tried to be summarised and some visions of CAx in 2007 were expressed. After the working days it was really surprising that there were no fundamental differences in the ideas of the future of CAx.

The universal topic of the workshop was the need for integration: of CAx-technology, of applications, and of co-operation enterprises – and additionally the interoperability of systems.

February 2000

Werner Dankwort, Josef Hoschek

Kaiserslautern / Darmstadt

Contents

CAD/CAM Data Interoperability Strategies

George Allen

Unigraphics Solutions, Cypress, California
allen@ugsolutions.com

1 Introduction

This document discusses some approaches to achieving data interoperability, which should be valuable to companies who are considering using two or more CAD/CAM/CAE systems together to address their product and process design needs. The remainder of the document is organised as follows:

- Chapter 2 provides background information on data types in modern CAD/CAM systems
- Chapter 3 describes some typical user scenarios and the interoperability needs that arise
- Chapter 4 is an overview of current data exchange technology
- Chapter 5 describes some ways to improve the convenience of data exchange
- Chapters 6-8 describe ways to improve the completeness of data exchange

1.1 Types of Interoperability

To place the issue of data interoperability in the proper perspective, we should point out that there are actually several different types of interoperability:

- Data interoperability, which is the subject of this document
- Platform interoperability — running on the same hardware
- User interface interoperability — using the same user interface concepts
- Custom application interoperability — supporting the same APIs for custom extensions
- Data management interoperability — single PDM managing files from multiple CAD systems

In each case, interoperability is valuable because it preserves some investment that has been made by the user, or leverages that investment more broadly by providing returns across the two systems. Specifically:

- Data interoperability preserves investment in creation of product & process data
- Platform interoperability preserves investment in hardware and support infrastructure
- User interface interoperability preserves investment in user training
- Custom application interoperability preserves investment in custom application programs

1.2 Attributes of Data Exchange Processes

For convenience, we think of data exchange processes as having two basic attributes, which we will refer to as convenience and completeness.

Completeness is a measure of how many different types of data can be exchanged. This might include issues such as:

- Associativity (preserve links during copy/edit)
- Preservation (no losses, upward/downward compatible)

- Reliability (error free, robust, precise)

Convenience is a measure of the ease of exchanging data. This includes

- Access (how to get the data)
- Ease of use (user-friendly, easy to operate)
- Performance (predictable, fast, economical)
- Programmability (customizable, adaptable, interfaces)

1.3 Summary of Conclusions

Geometry (curves and surfaces) is easy to exchange — see section 4.3.

B-rep solids can be exchanged using STEP, but better with a shared kernel like Parasolid. See section 6.3

Preserving parent-child relationships during model editing is complicated. Retaining associated data (like NC toolpaths) can be difficult, even within a single system. See section 2.9.

The techniques that are used to preserve parent-child relationships within a single system can be extended to work across multiple systems. See chapter 7. Persistent names are the key to success.

Exchange of feature and parameterisation data is difficult — see chapter 8.

2 Background Information on CAD System Data

This section provides background information for readers who are not familiar with the associative data structures used in modern CAD systems. Much of this material is derived from [1].

2.1 Intelligence, Parameterisation and Associativity

In modern CAD/CAM systems, most model objects are intelligent, or associative. A given object "remembers" that it was defined in terms of certain other objects, called its parents, and the child object updates automatically whenever its parents change. In this way, a network of hierarchical parent-child relationships can cause change to propagate through the model. The parent-child analogy can be extended liberally to provide a rich vocabulary for discussing inter-object relationships — we can use terms like grandparents, siblings, ancestors, descendants, and so on, with the obvious meanings. We can think of an intelligent object as consisting of two parts: a definition and a value. The definition records the defining data of the object plus the rule used to generate the object. The value is the result of using an evaluation or regeneration processor to "execute" the definition and obtain an answer.

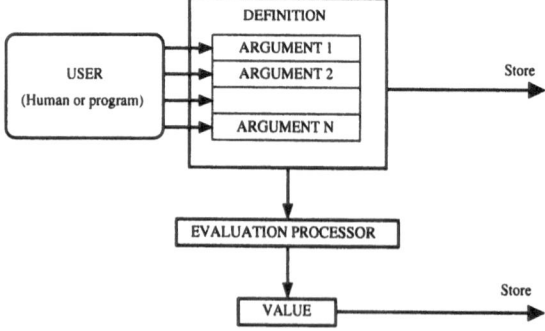

This is a somewhat abstract notion, so an example may help. Consider the case of a simple slab shape, as illustrated below.

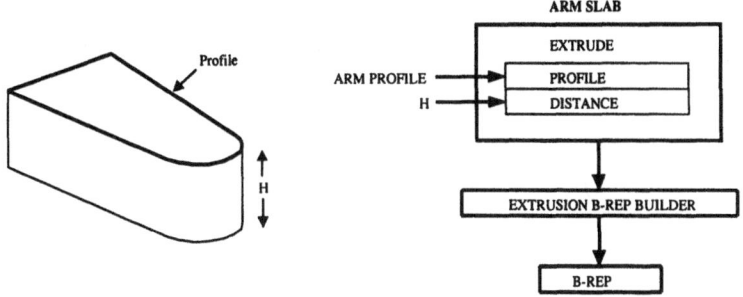

The parents of this slab body are of a set of curves describing its profile, and a numerical variable H specifying its height, and some positioning data (not shown in the picture). The result (the child) is a solid body consisting of twelve edges and six faces. We might say that we have established an associative parent-child relationship between the slab and its parents. If the value of H is changed, or if any of the curves in the profile are modified, then the system regenerates the slab definition to obtain a new boundary representation.

If the parents used in defining an object are numerical (like the height H in the slab example above), they are often referred to as the parameters of the object, and it is said to be parameterised or parametric.

2.2 Expressions

An expression is just a simple numerical variable. Its definition consists of a list of references to other expressions together with the formula (involving standard arithmetic operations, trigonometric functions, etc.) used to produce it. The value of an expression is just a real number.

2.3 Relative Positioning Constraints

Most systems use a variety of simple techniques for relative positioning of two bodies. In some cases, the bodies might represent piece parts in an assembly. In another scenario, one of the bodies is a "target" and the other body represents a feature that is to be attached to it (see below). The desired positioning is specified by imposing simple constraint relationships between geometric entities that belong to the two bodies. For example, one might stipulate that a cylindrical surface on one body is to be coaxial with a conical surface on the other body. One of the bodies (the parent) is considered fixed, and the system calculates a position for the other body (the child) in which the positioning constraints are satisfied. The constraints are "remembered" so that the positioning can be associative — if the parent body changes, the system can use the constraints to calculate a new position for the child body.

2.4 Sketches

A sketch is a collection of curves, together with dimensions and geometric constraints that determine their size and position. Sketches are often used in the definitions of extruded solids or solids of revolution, as we saw in the slab example earlier. An example of a simple sketch is shown below:

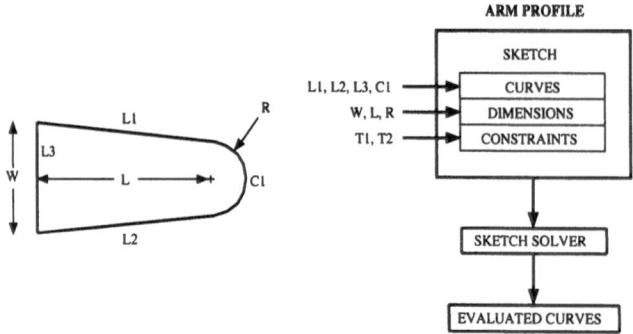

In this case the geometric constraints T1 and T2 indicate that the lines L1 and L2 are to be tangent to the circle C1. These geometric constraints may be specified explicitly by the user or may be inferred by the system as the curves are being constructed. The sketch solver modifies the curves so that they conform to the dimensions and constraints. Note that in general a sketch definition is declarative — we simply specify that certain constraints are to be satisfied, but we do not have a simple procedural recipe for producing a solution in a sequential series of steps. Solving sketches usually requires that the system somehow simultaneously solve a system of non-linear equations. There are many approaches to this problem, none of which is completely satisfactory at this point. See [6] and [7] for some partial solutions.

2.5 Features

A feature can be thought of as a parameterised collection of faces and edges that is to be attached to some target body. The feature might be positive (boss, pad, rib) or negative (hole, pocket, groove). Features can be loosely classified as:

- **Form Features**: Hole (simple, counterbored), pocket, boss, pad, etc.
- **Body Features**: Block, extrusion, free-form bodies
- **Operation Features**: Blend/round/chamfer, taper, offset, hollow, etc.
- **Enumerative Features**: Circular and rectangular arrays

In some systems, almost any associative object is referred to as a feature, but this obscures the term, so we choose to use it in the more specific sense described above.

The main value of features is in model editing — the user changes lower-level objects and relies on the system's associativity facilities to propagate this change throughout the rest of the design and any attached applications data. This technique affords tremendous editing leverage —global design changes can be introduced with very little user input. Furthermore, the parent-child relationships between features can serve to capture and enforce simple engineering knowledge.

Design features are very rarely used in applications because the features structure introduced during the design stage is unlikely to be the one needed for applications. The classical example is illustrated below: the designer thinks of this part as a single large pocket with a stiffening rib, but the manufacturing engineer sees it as two pockets side by side

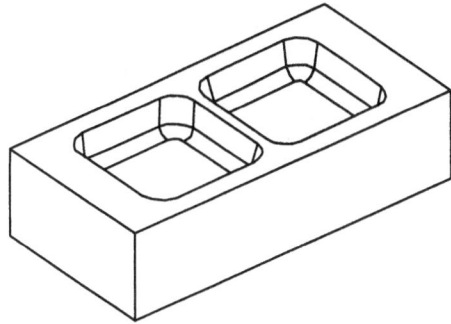

In some small vertically integrated companies, feature-based machining approaches are in use, but this is the exception, rather than the rule. Designers have been trained to design using features (such as removal volumes) that are useful for driving manufacturing algorithms, but they generally resist this, and it has few advantages, since most applications algorithms are not able to take advantage of feature information. Even in modern feature-based systems like Unigraphics or Pro/Engineer, there are only one or two application algorithms that are influenced by feature structure, and even these algorithms have "back-up" branches that are used when no feature data is present. In special environments with narrow and well-focused scope, custom applications can be built that can deliver enormous productivity improvements by leveraging feature information, but these applications are too specialised to be delivered as part of a general-purpose CAD/CAM system.

2.6   Feature Attachment

As noted in [3], the feature attachment operations in many systems are somewhat different from a conventional boolean union or subtraction. One advantage of this is that feature attachment operations sometimes produces more intuitive results. An example is shown below — we see an end view of a thin-walled pipe, and we wish to attach a cylindrical boss to the outside of it:

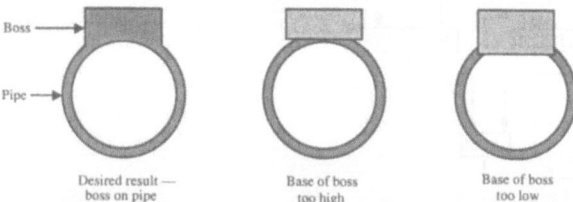

Boss

Pipe

Desired result —   Base of boss   Base of boss
boss on pipe   too high   too low

The naive approach, using conventional boolean operations, would be to unite a cylindrical solid with the pipe, but the illustrations show that this will not work. One can obtain the desired effect by using two boolean operations, but this is expensive. See [3] and [8] for further discussion of feature attachment operators. One of the problems with these newer feature attachment operations is that, unlike traditional boolean operations, their semantics are generally not well defined mathematically. In fact, he only real definition of the operation is the system code itself, and obviously this is different in different systems. Unfortunately this means that executing the same feature attachment operation in two different systems might well produce two completely different results. Therefore, if the same feature structure were exchanged between two systems and regenerated in each of them, the two answers would probably be different.

2.7   Applications Data

There will often be applications data that is attached to the model via associative parent-child relationships. For example, finite element meshes might be attached to solid bodies, or a toolpath might be attached to a face of the model as illustrated below

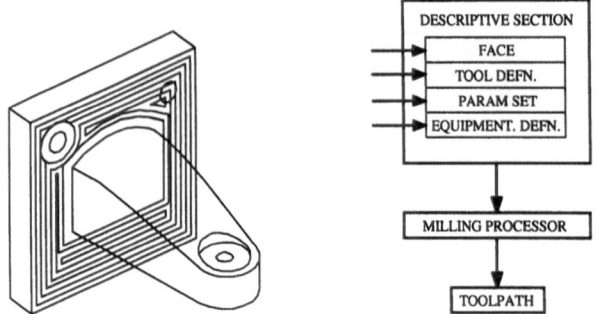

Again, the idea is that as the model changes, finite-element meshes, toolpaths, and other associated applications data will be updated accordingly.

2.8 The Model Graph

As we have seen above, there are many situations in which model objects are defined via associative references to other ones. The chains of references form a graph structure that we call the model graph. In [2] this structure is referred to as an E-rep (editable representation). It is quite similar to a traditional CSG graph — the terminal node represents the final model, and internal nodes represent its constituents and various intermediate stages in its construction.

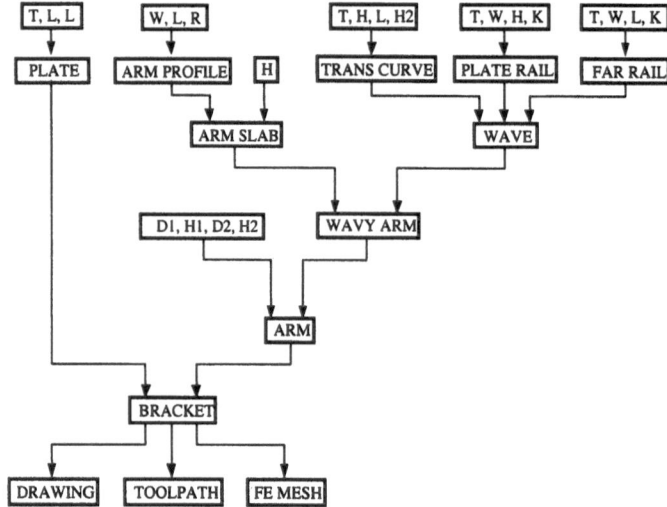

The model graph plays a crucial role in editing functions. When the user changes some object, the parent-child links in the model graph are used to decide which other objects need to be regenerated to reflect the change. For example, if the user changes the value of some expression variable, we may have to calculate new values for several other expressions that depend on it, which may cause the sizes and/or locations of several features to change, and so on. This propagation of change can be extremely powerful, and users seem to grasp it quite readily. We generally explain the process by telling users that it is analogous to the behaviour they have seen in spreadsheets.

Spreadsheet	Associative Model
A3 = B1 + B2 ↑ ↑ ↑ Child Parents	**Solid3 = Solid1 + Solid2** ↑ ↑ ↑ Child Parents
R5 = Avg(Q1..Q7) ↑ ↑ Child Parents	**Surf5 = SurfaceThruCurves(C1..C7)** ↑ ↑ Child Parents
C2 = 0.25*C1 **A2 = B2 + C2**	**Diameter = 0.25*Width** **H3 = Hole(Diameter, Depth, ...)** ↑ ↑ Parameters

2.9 Maintaining Parent-Child Relationships

As a model is edited, correctly maintaining associative parent-child relationships can sometimes be difficult. Consider the rather extreme example shown below. The user has created an N-sided prism, with N initially equal to 4, and then selected a face to be machined, thereby creating an NC toolpath on one of the faces. This is shown in the illustration on the left, with the toolpath depicted by a black rectangle.

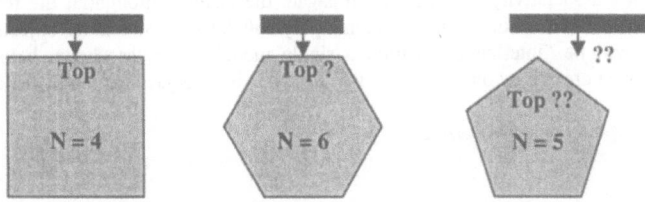

Now the user changes the value of N from 4 to 6, which will cause the model to be regenerated. During regeneration, the system has to decide which of the faces of the 6-sided prism should have the toolpath attached — in other words, which face should be used as the parent of the toolpath. The problem is that the user did not explicitly express any design intent when he originally selected the face to be machined. Perhaps he meant that the "top" face was to machined, in which case we might be able to regenerate the model in the case N=6, as shown in the center illustration. However, we still have a problem if the user changes the value of N to 5, since in this case there is no obvious "top" face.

Usually, maintaining parent-child relationships is not as difficult as in the example outlined above. The operation that creates a body "names" its faces and edges, and these names are used when attaching children. So, in our example, the operation that creates the prism might produce faces named "top", "left", "right", and so on, and the toolpath will remember that it is supposed to be attached to the "top" face. The prism-making algorithm must guarantee that the names will be persistent, in the sense that faces having analogous roles will be given the same name, as far as possible, every time the prism is regenerated. In the prism case, this is quite difficult, but clearly it is fairly easy to devise a persistent naming scheme for the faces and edges of an object like a block, since the number of faces and edges never varies. The persistent naming issue is at the very core of every parametric or associative system — if modeling functions cannot assign names to their results in a persistent way, then there is no reliable way to attach children. When the model is regenerated, children will be unable to locate their parents in the new version of the model, so they will become unattached orphans or will disappear completely.

Note that associative parent-child relationships are even more difficult to maintain if the user changes the model by deleting and recreating objects, rather than by modifying them in place. Consider our prism example again, suppose as before that there is some application data (like an NC toolpath) attached to the "top" face of a four-sided prism. If the user simply edits the prism, changing it into a six-sided one, there is some chance that the techniques outlined above will succeed in keeping the toolpath attached to the face correctly. On the other hand, suppose the user deletes the four-sided prism, and then creates a new six-sided one. At best, the attached toolpath will be "orphaned" when its parent face is deleted. The system has no way of knowing that the new six-sided prism is actually meant to be a replacement for the original four-sided one, so it will be difficult for the toolpath to be reattached correctly.

Repeated import of models from another system is typically a delete-and-recreate operation with a very coarse level of granularity, so it causes particularly severe problems in maintaining parent-child relationships. This is discussed further in chapter 7.

3   User Scenarios for Interoperability

Different levels of interoperability are required at different stages of the product development process. Several different scenarios are outlined below, along with their interoperability requirements.

3.1   Interleaved Collaborative Product Design

If two groups or individuals are collaborating closely on a product design project, it is likely that each of them will want to make changes to the same model at some stage. In most contemporary systems, the most effective way of changing a design is to edit lower-level objects (typically design features) and rely on the system's associativity facilities to propagate the change throughout the rest of the design and any attached applications data. Therefore, in highly collaborative situations, objects created by the two designers will be "interleaved" within a single model graph, as shown below. Here ▬▬ denotes an object created by one designer and ▭▭ is an object created by the other.

This sort of "interleaved" collaborative design requires the ability to exchange design features and parameterisation structures, which is extremely difficult, for the reasons described in section 8.1. If two designers want to make interleaved modifications to the same model graph, then, given the current state of technology, they must use the same CAD system.

An alternative approach to collaborative design is to provide effective editing functions that do not depend on the model graph structure. Functions can be implemented to manipulate individual faces or groups of faces, regardless of how they were constructed. The CoCreate system implements this approach exclusively, and Unigraphics also provides some support for this form of editing.

3.2 Non-Interleaved Collaborative Product Design

The idea here is that one user is constructing some portion of the model, and another user is constructing another portion that depends on the first. Often the two portions of the design are different piece parts, although this is not necessarily the case. The situation is illustrated below.

Examples of this sort of scenario are common in many industries. For example, in the aerospace industry, design of structural parts depends on the design of loft surfaces, and in auto body work, design of inner panels depends on the design of outer panels. From a technical point of view, the first designer is working on a "parent" and the second designer is working on a "child". This type of design scenario may be concurrent, rather than purely sequential. In fact, it will often be useful to begin the "child" portion of the design before the parent is complete. Whenever a new version of the parent design emerges, the child portion of the design can be reattached to it and the child will automatically adapt accordingly.

This sort of design scenario does not require exchange of feature information. From the point of view of the child, the parent design is simply a collection of faces and edges to which new objects can be attached. The second designer does not need to modify the parent design, and in fact is often expressly forbidden from doing so. Designers of airframe structural parts are certainly not permitted to change loft surfaces, for example. However, as new versions of the parent design are released, it is important to have some way to ensure that objects in the child design can remain attached properly. See chapter 7 for further discussion of ways to preserve parent-child relationships.

3.3 Tooling Design

Tooling design is one specific example of non-interleaved collaborative design, as described above. In this particular case, the "parent" is the part design, and the "child" is the tooling needed to produce it. The tool designer ideally will not need to modify the product design, so he does not need access to its feature structure. There are a few cases where tool designers will actually want to modify part designs; for example they need to add draft to improve manufacturability, and the designer may not be available to do this. This can be done by adding new "draft" features to the product model. An alternative approach is to actually modify the features and parameters of the product model to produce the desired draft, but this will require access to the feature structure and will not be possible unless the tool designer uses the same system as the product designer. Again, it is likely that development of tooling will begin before product design is complete, and it is important to be able to keep these up to date with a minimum of effort as the product design evolves. The tooling model will typically be linked to the solid bodies in the product model, so associative updates can occur automatically though the mechanisms described in chapter 7. In fact parent-child relationships where the parents are entire bodies (rather than faces or edges) are generally fairly easy to maintain during editing.

3.4   Generation of NC Instructions

Generation of NC instructions is again similar to non-interleaved collaborative design, as described above. In this particular case, the "parent" is the product or tool design, and the "child" is the NC program needed to produce it. Development of NC programs does not generally require access to the feature stucture of the object being machined. Again, it is likely that development of tooling and NC programs will begin before product design is complete, and it is important to be able to keep these up to date with a minimum of effort as the product design evolves. The NC programs will typically be linked to the faces and edges of the product model, so again associative updates can occur automatically though the mechanisms described in chapter 7.

.3.5   Non-Associative Re-Use of Design Data

In some uses of design data, the data is merely read — no new data is created or attached to the "parent" design, so no new associative relationships are established. One good example is digital mockup or fly-through applications. In fact, digital mockup is a particularly easy application since it requires only facet and assembly structure data to be exchanged..

3.6   Process Analysis

From the examples described above, we can see that careful analysis is required to understand what type of interoperability is required in a given product development scenario. Among other things, we need to understand

- The type of data needed at the destination
- How the data will be used
- How often transfers will occur
- The amount of data to be transferred
- Whether new objects will be associated with the transferred data
- Whether transfers will be repeated

Once these requirements are properly understood, best practices can be developed, documented and institutionalised. In some situations, it may be advisable to capture these best practices in custom programs, so that process consistency is ensured.

4   Data Exchange Technology Overview

The types of data we might wish to exchange can be roughly categorised as follows::

- Simple graphics data
- Facet data
- Geometry
- Boundary representation topology
- Drafting symbols
- Data organisation
- Associative relationships
- Feature and parameterisation data

We consider each of these in the following paragraphs.

4.1   Simple Graphics Data

Vector-oriented 2D graphics formats are adequate for exchanging graphical representations of engineering drawings and other types of illustrations. Simple graphics data is easy to exchange using well-known standards such as CGM, SVF, Windows metafiles, PostScript, and various bitmap formats. The simplicity and broad acceptance of 2D graphical formats explains the ease with which images and illustrations can be transferred between PC applications like MS Word, PowerPoint, and Excel.

4.2 Facets

Facet data is often used for high-end graphics applications and for spatial studies of fit, clearance, and accessibility. Many systems for visualization and analysis of large assemblies are driven entirely by facet data. The number of facets used can be adjusted to make sensible tradeoffs between accuracy and performance. This type of data is generally easy to exchange. Common formats include VRML, SLA (stereolithography files) and various graphics file formats such as Inventor.

4.3 Geometry

Geometry data (curves and non-trimmed surfaces) is central to many engineering and manufacturing applications. The most basic building blocks for representing shape information are geometric entities — points, curves, and surfaces. Points are so simple that they do not warrant any further discussion. The curve types generally available are:

- Line (i.e. straight line)
- Circle
- Conic section curves (ellipse, parabola, hyperbola)
- A spline curve, usually a non-uniform rational b-spline (NURB) curve

Similarly, the surface types typically available are

- Plane
- Circular cylinder
- Circular cone
- Sphere
- Torus
- Surface of revolution
- Extruded surface
- Offset surface
- Some sort of free-form surface, usually a NURB surface

This is roughly the set of geometric entities specified in standards like IGES and STEP, and (more importantly) is roughly the set of geometric entities supported by most popular systems. Simple engineering geometry (lines, circles, planes, cylinders) has been standardised for hundreds of years, and is easy to exchange. In recent years, with the almost universal adoption of NURBs, free-form geometry has also become easy to exchange, although there are sometimes problems because different systems have different limits on the mathematical degree of NURBS curves and surfaces. Because of this commonality, of geometry forms, exchanging curve and surface data between systems generally works fairly well. Conversely, market demand for good basic data exchange facilities and "open" systems deters vendors from deviating too far from this data model, so it is likely to remain quite stable in years to come. Geometry data can be exchanged very effectively using standards such as IGES and STEP. STEP is somewhat preferable because the geometric objects are more rigorously defined.

4.4 Boundary Representation Topology

Most systems use some form of boundary representation (b-rep) in which curves and surfaces are combined together to describe (solid or sheet) bodies. The basic topological entities are vertices, edges, loops, faces, shells and bodies. They record various types of adjacency or connectivity relationships, as illustrated in the figure below.

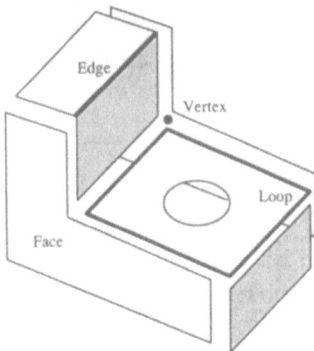

So, for example, we might store information indicating that two faces are adjacent in the sense that they meet along some common shared edge. Note that this is logical information, and in fact the two faces are typically allowed to have a small gap between them, provided this is less than some tolerance. Problems arise when exchanging topology information because different systems use different tolerances when deciding whether or not two faces are really adjacent. The gap between faces might be within the tolerances of one system, but outside the tolerances of another, thereby contradicting the topological information indicating that the two faces are adjacent. In other words, we arrive in a situation where the geometric information and the topological information are inconsistent. Parasolid creates models to very tight tolerances, so exporting from Parasolid to another system typically works very well. To import models from a less precise system, Parasolid has a special "tolerant modeling" facility that allows it to handle models that are less precise than the ones it typically produces itself. The tolerant modeling innovation helps considerably, but topology problems still arise when transferring solid models between systems using standards such as STEP and IGES. These problems can be eliminated by using the same modeling kernel in the systems that need to exchange solid models. For example, exchange of solid models between Unigraphics and Solid Edge works flawlessly because they both use the Parasolid kernel.

4.5 Drafting Annotations & Symbols

Drafting annotations (notes, labels, dimensions, crosshatching) are actually quite difficult to exchange effectively. The problem is that they are represented in quite different ways on different systems. In particular, many layout problems arise from font differences. Both IGES and STEP have facilities for exchanging drafting annotations, but it is surprisingly difficult to transfer drawings between systems and have them look identical afterwards. Of course, if identical appearance is all that's required, simple graphical formats (CGM, PostScript) can be used instead, though this loses all the structure and "intelligence" in the drawing and makes it fairly difficult to edit in the receiving system.

4.6 Data Organisation

Data organization structures include groups, layers, categories, and assembly structures. Provided the participating systems have equivalent capability for representing organisational structure, exchange should work smoothly, since this is logical data that is not subject to precision problems that occur with floating point data. Most contemporary systems have roughly equivalent facilities for representing layer, group, and assembly structures, and these structures are well supported in IGES and STEP AP214.

4.7 Associative Relationships

It is often useful to establish associative parent-child relationships between objects owned by the two systems. This is difficult to achieve if the two applications share data via file transfer, rather than by data file unification. The problem is that file transfer is a "delete and recreate" operation with a very coarse level of granularity, and it is difficult to keep child objects properly attached to parents who are continually dying and being reborn en masse. The situation can be improved by using incremental file transfers (transferring only the objects that have changed) or by requiring the sending system to assign persistent identifiers that can be used to track objects as they are deleted and recreated. This topic is discussed in detail in chapter 7.

4.8 Feature and Parameterisation Data

Feature and parameterisation information is difficult to exchange, but fortunately this does not have a great impact on applications effectiveness. These issues are discussed in detail in section 2.5 and chapter 8.

5 Improving Convenience

This section discusses some techniques that might be used to improve the convenience of data exchange processes.

5.1 Concatenated Transfers

Data exchange based on standards is typically a two step process. Data is exported from the originating system in a standard format, and then imported into the target system. If these two steps can be concatenated into a single operation, this obviously improves convenience. Also, if the target system is known at the time the data is exported from the originating system, then various parameters that control the process can be set accordingly. For example, if the Catia system knew that it was exporting a STEP file destined for Unigraphics, then zero-length curves can be eliminated, tolerances can be set appropriately, and so on. The operation could even be presented in the Catia user interface as an "Export to Unigraphics" function that embodies many of the best practices involved in this form of exchange.

5.2 Bundled Transfers

In some cases, for two given systems, there are two separate data exchange paths that have complementary capabilities. Suppose for example that we want to transfer a file from Solid Edge to Unigraphics. Using a Parasolid transmit file provides flawless exchange of solid b-rep data, but would not transfer annotations and drafting symbols. On the other hand, IGES or STEP would handle drafting annotations, but might have some trouble with solids. The net result is that two separate data exchange mechanisms need to be used to achieve optimum completeness. This is inconvenient, and also fails to handle any associative relationships between drafting annotations and edges of the solid model. The data exchange is certainly more convenient if the two data exchange channels are "bundled" into one. For example, the Parasolid model could be embedded as a binary object in a STEP file that also includes a wide variety of other data (such as drafting annotations). This improves convenience, and also makes it easier to preserve associative relationships.

5.3 Selective "Pulling" of Data

The typical two step translation process described above requires the user of the sending system to "push" data out in a standard format, which can then be read into the target system. This arrangement sometimes causes logistic problems if the user of the sending system is not available when the data is needed. Also, the user of the sending system often has difficulty understanding which data is required, and decides to export all of it, thereby overwhelming the person on the receiving end. Both problems can be solving by providing a function that allows the user of the destination system to selectively "pull" data from the sending system. A mechanism like this has been implemented at General Motors to allow Unigraphics users to pull data from GM's internal CGS system, and this has been very well received.

5.4   User Interface Conveniences

The convenience of data exchange processes can be improved significantly by the use of appropriate user interface techniques. A good user interface can make the translation process more accessible, easier to understand, and more efficient. For example, data translation processes can be hidden with the familiar "Open" and Save As" commands, so that the user doesn't even need to think about them. Alternatively, in the Windows environment, translation can be set up as one of the actions on the context menus that appears when the user right-clicks on a file in Explorer. Another idea is to present translation of selected objects as a Cut & Paste operation, which is familiar to all users of popular software products nowadays. The Cute & Paste approach also has the advantage that it allows selective "pulling" of data, as described in the previous section.

6   Improving Completeness

This section discusses some techniques that might be used to improve the completeness of data exchange processes. Further discussion of specific techniques for associative relationships and features can also be found in chapter 7 and chapter 8 respectively.

6.1   Process Improvements

Significant improvements in data exchange effectiveness can be achieved simply by developing, documenting and institutionalising best practices. This is likely to involve recommended modeling practices, data preparation techniques, choice of the appropriate exchange mechanism (where there are choices), and validation of the process.

6.2   Improve Scope of Standards

Sometimes data exchange effectiveness is impaired because of the nature of the exchange standards being used. There are two sorts of problems, both of which involve a mis-match between the representational scope of the standard and the systems being used. In one scenario, there might be an object type that is present in both systems but can not be represented in the standard, and which therefore can not be exchanged. Some standards are open-ended, which allows co-operating vendors to exchange object using a mutually agreed "private" representation that is not part of the formal standard.

On the other hand there are cases where exchange standards are too liberal and all-encompassing. Adding an object type that is supported by only a few systems makes it easy for those few vendors to write preprocessors, but makes it difficult for everyone else to write postprocessors. This increases the total amount to software that has to be developed, and slows down the progress of data exchange technology.

6.3   Sharing a Common Kernel Modeler

As indicated elsewhere, use of a common kernel modeler can dramatically improve the ability to exchange geometry and b-rep topology information between systems. Thus, for example, Unigraphics and Solid Edge can exchange models very effectively because they both employ the Parasolid kernel. To foster this form of data exchange, Unigraphics Solutions licenses Parasolid to a variety of different companies who produce engineering and manufacturing applications software. Our primary goal is to encourage the development of applications that can interoperate seamlessly with Unigraphics, either by sophisticated end users or by third party software vendors, but we are also willing to supply Parasolid to vendors of full-function CAD/CAM/CAE systems that could potentially compete with Unigraphics. There are many successful users of Parasolid; some notable examples are MSC (CAE software), Concentra (Knowledge-based engineering software), Deneb (manufacturing simulation), Mitsubishi Motors (internal CAD system), and SolidWorks.

Data exchange via Parasolid could be improved by further standardising the way it is used in different systems. Everyone uses Parasolid in the same way to represent solid bodies, but there are some discrepancies in the way various systems use Parasolid to represent assemblies, non-geometric attributes, and isolated curves. Improved standardisation in these areas could improve data exchange effectiveness.

In addition, there have been data exchange problems in the past because, as a result of their release schedules, different systems have been using different versions of Parasolid for a while. This problem has been resolved because a given version of Parasolid can now save data in older formats.

6.4 Full Coverage Read/Write APIs

Unigraphics has a rich application programming interface (API) called UG/Open that provides, among other things, complete read/write access to all model data. This API is actually used by our internal development groups to write data exchange software, but it is also available to others outside Unigraphics Solutions. This means that sophisticated end user companies who have their own programming staffs can write customised translators if they wish. Our Parasolid kernel also has a complete read/write API, and several end users and third parties have used this to write data exchange software. Commercial translators based on Parasolid have been produced by Theorem Solutions and Armonicos, for example.

6.5 Publicizing File Formats

Providing APIs, as described above, is one way to open the data exchange field to users and third parties. Another useful practice is publishing of file formats. Some groups writing translators and other software prefer to read files directly, rather than using an API supplied by the system vendor. This is sometimes less expensive because it does not incur any licensing costs, and may produce a software package with a smaller foot-print, since it more tailored to the task at hand. On the other hand, use a vendor-supplied API provides more stability and version independence. To provide application developers with a choice between the two approaches, Unigraphics Solutions recently publicised the format of the Parasolid XMT format. The ACIS SAT file format has been public for several years.

7   Preserving Associative Relationships

This section discusses the specific issue of preserving associative parent-child relationships during data exchange processes. Specifically, we envision a situation involving tool design or some other associative re-use of design data in the receiving system, as described in the scenarios above. To make the discussion more concrete, we consider a case that actually occurred at one of our customer sites: the originating system is Catia, the target system is Unigraphics, and NC toolpaths are the "child" objects being associated with the transferred model data after it arrives in Unigraphics. The problem is that we would like the toolpaths to remain properly attached to the geometry when new versions of the design are imported from Catia.

In the descriptions below, we have discussed a scenario in which toolpaths are attached to faces of solid models. In some systems, toolpaths are actually attached to solid bodies that represent removal volumes, but the basic principles and the approach are much the same.

As indicated in section 2.9, repeated import of models from another system is typically a "delete-and-recreate" operation with a very coarse level of granularity, and it is difficult to keep child objects properly attached to parents who are continually dying and being reborn en masse. This section outlines some ways to address this issue.

7.1 Incremental Transfers

If an object in the design has not changed from one version to the next, then it is best not to re-transfer this object. That way, the old version remains intact in the target system (Unigraphics) and so all its children (toolpaths and other associated data) are undisturbed. In other words, we would like the data transfer to be incremental. There are several ways to achieve this. One approach is to implement some mechanism for recording data transfer operations and model changes in the originating system. Then, whenever a data transfer function is requested, the system would simply export only those objects that have changed since the last transfer to the same destination. Another approach is filtering during import into the receiving system. The receiving system would scan the incoming objects and compare them with objects already present in the model. If an incoming object is found to be identical to one that is already present in the target model, it should be skipped, so the original object remains intact along with all of its descendants. More generally, we should try to detect new objects that correspond to existing ones, rather than just being identical. If we conclude that an incoming object is simply a new version of an existing one, then it should replace the original, and all children of the original should be reattached to the new version.

The situation is illustrated below. The design is a simple block shape, and toolpaths (depicted by black rectangles) are attached to its left, top, and right faces. Then the design is changed — in revision 2 the right face is moved to the right, which means that the top and bottom faces are extended.

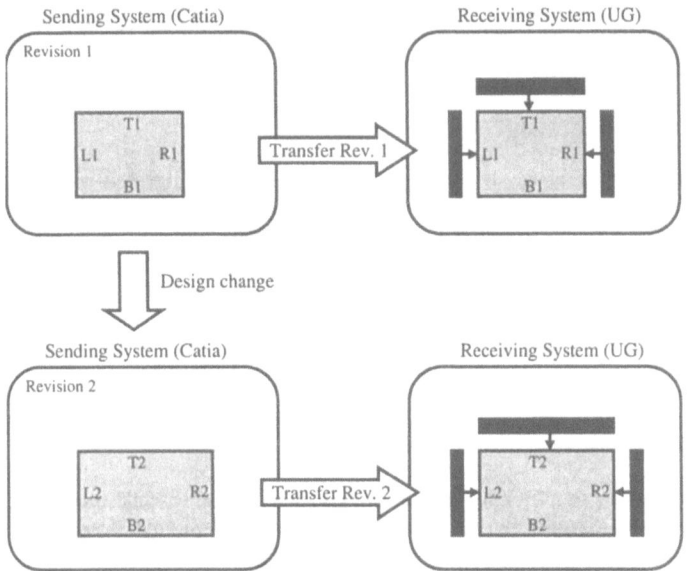

The second time the design is transferred to Unigraphics, the following should happen:

- The left face (L2) should not be imported into the UG model, since an identical face (L1) already exists
- The right face (R1) should be replaced by face R2, and the attached toolpath should be regenerated

The top face (T1) should be replaced by face T2, and the attached toolpath should be regenerated

Detecting the fact that L2 is identical to L1 is fairly easy using simple geometric comparisons. It is more difficult to decide that T2 and R2 are just new versions of T1 and R1 respectively, but we have prototyped algorithms based on heuristics that work much of the time. For example, we might conclude that R2 is just a moved version of R1 because it has identical area, perimeter, moments, geometric form, and so on. Similarly, we might decide that T2 is a new version of T1 because it lies in the same surface and three of its four edges are very similar. Once we decide that T2 is a new version of T1, we can delete T1, import T2, re-attach the toolpath to T2, and regenerate it. The tool motions will be different, but tool data, cutting parameters, and other information will be retained, so the user does not need to re-enter these.

7.2 Manual Reattachment Operations

In typical situations, the heuristic algorithms described in the previous section will preserve most of the child objects of a modified design, but they will never work perfectly. From time to time, there will be cases where the algorithm cannot detect the correspondence between old and new versions of a particular face, so both versions will end up in the model, with the toolpath attached to the old one. In these cases, the user must delete the old version of the face and manually re-attach the toolpath to the new version. This sort of breakage of parent-child relationships sometimes happens even within a single application, so most contemporary systems have a variety of functions for repairing the relationships.

7.3 Persistent Identifiers

As outlined in section 2.9, parent-child relationships within a single system are preserved during editing by the use of persistent identifiers that indicate when new objects should be considered merely as new versions of old ones. For example, in order for associativity to work within Catia, the right-hand face of our block shape must be continue to have the same persistent name (like "right") regardless of how the user moves it around. Then, child objects in the Catia model can simply remember that they are to remain attached to the face named "right", and everything works. It is possible to extend this scheme across multiple systems by exporting these persistent names via the chosen data exchange mechanism. In fact, the STEP format has facilities for transferring persistent names, and it should be relatively easy to make use of this mechanism. The process would work as follows:

- In the transfer of revision 1, face R1 would be named "right" in the STEP file output from Catia
- This name "right" would also be attached to the corresponding face in the UG model.
- After the design change, if Catia associativity works correctly, face R2 would now be called "right"
- In the transfer of revision 2, face R2 would now be named "right" in the STEP file output from Catia
- The matching name "right" would tell us that R2 is just a new version of R1
- When UG imports the second STEP file, R2 would replace R1 in the UG model
- All children of R1(like the toolpath) would be "reparented" and would become children of R2 instead

The attached toolpath (and any other children) would be regenerated.

This scheme works much better than the heuristic comparison scheme described above. It does not force the receiving system to "guess" that the imported objects are intended to be new versions of old ones — this intention is made explicit by the persistent names, just as it would be within a single system. However, this scheme only works if the originating system has persistent names that can be exported into STEP files.

8 Exchanging Feature & Parameterisation Data

One of the most difficult challenges in data exchange is handling feature and parameterisation data. This section explains why the problem is fundamentally difficult, and explores some possible approaches.

8.1 Problems Exchanging Features

As indicated in section 2.6, the semantics of feature attachment operations are not formally defined in any rigorous mathematical way, so implementations differ across different systems. For example, the illustrations below show the possible results that might be obtained from three algorithms that claim to attach a circular boss to the top of a block. The situation is a little unusual since the boss hangs over the edge of the block, but is still typical of the types of feature attachment operations performed every day in contemporary systems.

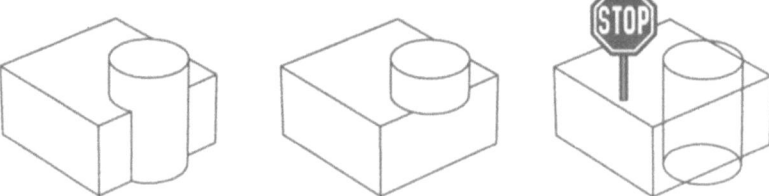

We see that the results are entirely different, and in some systems this type of "overhanging" boss is actually regarded as an error. So if we take a model graph that defines the object above, and exchange it between three systems, then, after regeneration, we might well get three different results. The fundamental problem here is that the meaning of the model (the result of regeneration) is actually dependent on the code in the system in which the regeneration is performed. Therefore, in order for a given model to have the same result in different systems, these systems must use the same regeneration algorithms. Since these algorithms are so subtle, and their results are not rigorously defined, in practice this means that the different systems must actually use the same regeneration code, and this poses some fundamental problems.

Experience with feature-based CAD systems like Unigraphics and Pro/Engineer over the last few years has shown that it takes considerable effort to exchange features even between two consecutive versions of the same system. A small change in an algorithm can lead to failures in features that regenerated successfully in a previous version. The only way to solve this problem seems to be to associate a specific version of the regeneration code with each individual feature, so again we see that identical regeneration results can only be obtained by using identical code. The sections below suggest some approaches to exchanging feature and parameterisation data, but none of the ideas is well developed at this point, and it is likely to be quite some time before this type of data can be exchanged successfully.

8.2 Standardising Features

There are a few on-going attempts to standardise features across CAD systems. Notable ones are STEP AP224 and the OCAI initiative. Current systems do not have anywhere near the same set of feature types, and, as we saw above, even systems that have the same feature types might have very different semantics.

8.3 Client/Server Approaches

Another idea is to use a client/server approach to make one system's regeneration algorithms available for use by another system. This is somewhat in the spirit of the OLE for Design & Modeling (OLE4DM) initiative. These ideas have already been used successfully to generate NC toolpaths and finite element meshes on models composed of mixed data from several systems. The current OLE4DM API only covers simple geometric and topological inquiries, but it is conceivable that it could be extended to provide feature regeneration services.

8.4 Object-Oriented Approaches

An object-oriented purist would say that features should be regarded as objects that carry their regeneration methods encapsulated within them. Then, when features are exchanged between systems, the regeneration code travels along with them, and is available for use by the receiving system. This idea industrial design fine in principle, but it requires an ability to dynamically transfer code between systems and have it used during regeneration. Perhaps a language like Java could provide the portable executable code that would be needed to make this possible, but this idea is only in the research stage, and no current system is architected this way.

9   References

[1] George Allen, "Tolerances and Assemblies in CAD/CAM Systems", Journal of Mfg. Systems, 1995

[2] C. M. Hoffman and R. Juan, "EREP: An Editable High-Level Representation for Geometric Design and Analysis", Technical Report CER 92-24, Department of Computer Science, Purdue University, 1992.

[3] C. M. Hoffman, "On the Semantics of Generative Model Representations", Technical Report CER 93-07, Department of Computer Science, Purdue University, 1993.

[4] C. M. Hoffman, and X. Chen, "Towards Feature Attachment", *Computer-Aided Design*, Vol. 27, 1995, pp. 695-702.

[5] C. M. Hoffman, V. Capoyleas and X. Chen, "Generic Naming in Generative, Constraint-Based Design", *Computer-Aided Design*, Vol. 28, 1996, pp. 17-26.

[6] R. A. Light and D. Gossard, "Modification of Geometric Models Through Variational Geometry", *Computer-Aided Design*, Vol. 14, No. 4 (January 1982), pp. 209-215.

[7] J. C. Owen, "Algebraic Solution for Geometry from Dimensional Constraints", *Proceedings First ACM/IEEE Symposium on Solid Modeling and Applications*, Austin, Texas, 1991.

[8] Mervi Ranta, Masatomo Inui, Fumihiko Kimura, Martti Mantyla, "Cut and Paste Based Modeling with Boundary Features", *Proceedings Second ACM/IEEE Symposium on Solid Modeling and Applications*, Montreal, Canada, 1993.

[9] K. J. Weiler, "Topological Structures for Geometric Modeling", Ph. D. Thesis, Rensselaer Polytechnic Institute, 1986.

Product Data Technology – A Basis for Virtual Product Development

Reiner Anderl

Technische Universität Darmstadt, Fachgebiet Datenverarbeitung in der Konstruktion (DiK)
anderl@dik.tu-darmstadt.de

Abstract: Information and communication technology is getting an increasing importance in the product development and design process and it is strongly influencing the competitiveness of industrial enterprises. Significant technical progress has been achieved by introducing modern CAX-systems and by establishing and using internet, intranets and extranets. Information and communication technology therefore is changing business processes and working cultures by moving from paper based and document driven development and design processes to development and design processes based on digital product and process models. This is called Virtual Product Development. This contribution presents major concepts of Virtual Product Development and explains the role of product data technology and its implementation as a basic platform for Virtual Product Development.

1 Introduction

The efficient application of modern information and communication technology in industrial enterprises requires new strategies, appropriate organisation and working styles. A design methodology, integrating the capabilities of modern information and communication technologies is required [1]. Based on the experience resulting from the scientific education of mechanical engineers, five strategic topics are of importance:

1. Thinking in processes,
2. Development and design based on 3D-CAD,
3. Documents to be understood as presentation of digital models,
4. Digital information available and accessible by any authorised user,
5. Use of product data management integrated with CAX-systems

A reflection of the application of CAX-systems in the development and design process shows mainly four stages which are illustrated in figure 1.

In the first stage, no CAX-systems are used. Product development and design is being performed on paper and the resulting documents are used for developing, for documenting and for communicating the technical solution.

In the second stage CAX-systems have been introduced in terms of islands of solution. CAX-systems supported engineers in developing their technical solutions but the documentation and communication has still been performed through documents on paper.

In the third stage interfaces for data exchange have been introduced. These interfaces are standardized interfaces like IGES (Initial Graphics Exchange Specification), SET (Standard d'Échange et de Transfert), VDAFS (Flächenschnittstelle des Verbandes der deutschen Automobilindustrie) and STEP (Standard for the Exchange of Product Model Data, ISO 10303) or industry standards like DXF (Data Exchange Format). Such interfaces enabled digital communication. First this communication has been performed as data exchange between CAD-systems but more and more other CAX-systems have been interfaced, too. The documentation however has still been performed on paper documents such as technical drawings and part lists or bills of material.

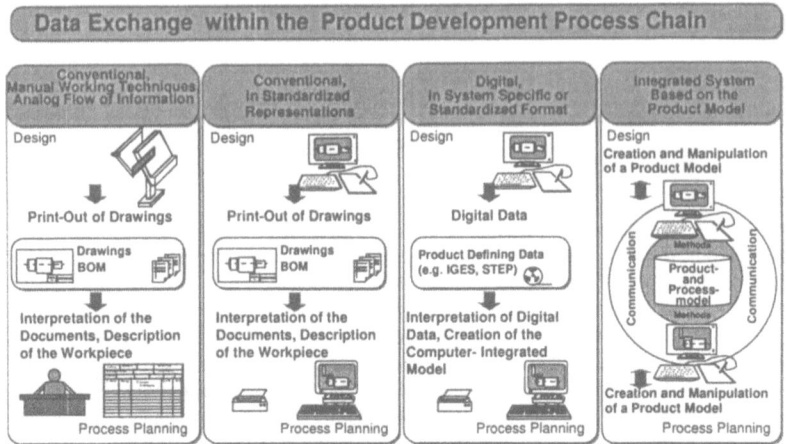

Figure 1: Stages of CAX-application in the development and design process

The fourth stage now aims at an integration based on databases, managed through PDM-systems and performing an integration for the required CAX-systems. Resulting from an increasing use of CAX-systems more and more digital models are being used and it is becoming more and more difficult to use documents on paper to represent the information and the knowledge which is represented in the digital models. Another strong issue is the requirement for a coherent digital model covering the information of the CAX-systems. This requirement leads to the so called integrated product model.

2 Concepts for Virtual Product Development

In Virtual Product Development the product development process is managed through PDM-systems supporting concurrent design and simultaneous engineering and using powerful application software systems for to engineer, to develop and to design innovative and multidisciplinary products. The main objective is to describe a product as a digital product model and to analyse and to simulate the products´ behaviour.

The management of the product development process is aiming at a fast time to production while assuring high development quality and keeping cost objectives. Another goal is the ability of the creation of variants and alternatives to enhance the attractiveness of the product.

Concurrent design and simultaneous engineering are management methods for controlling a complex development process which is being changed from a sequential process to a simultaneous one. This implies the overlap of development activities as shown in figure 2, often even involving cooperating enterprises.

For the management of Concurrent Design and Simultaneous Engineering in a Virtual Product Development process, its organisation is of crucial importance. This implies project management, release control, change management, document and data management as well as product configuration management. These management issues are provided by the PDM-system which is customised according to the requirements of the enterprise.

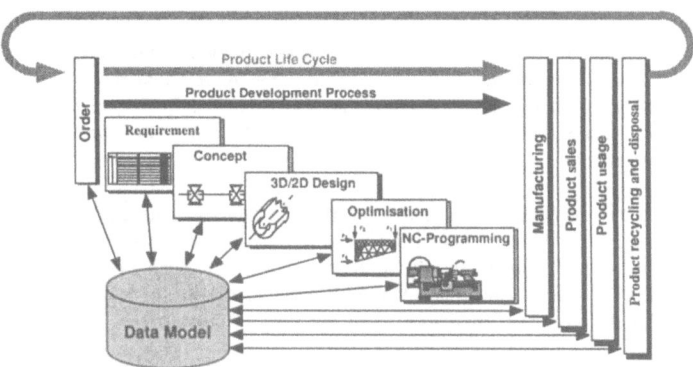

Figure 2: Concurrent Design and Simultaneous Engineering in Virtual Product Development

A number of powerful application software systems are being used for Virtual Product Development [2],[3]. These application software systems require a change in the working culture, in particular the move from a document driven working stile to a working style based on one or more digital models ideally based on the product model. That is why this product development process is called Virtual Product Development.

In Virtual Product Development, the product development process uses CAD-systems in early design phases for preliminary design and in particular 3D-CAD-systems for shape design. Besides the geometric product representation and the representation of the product structure, 3D-CAD-systems also cover product data such as material data, technology data and so called features (figure 3).

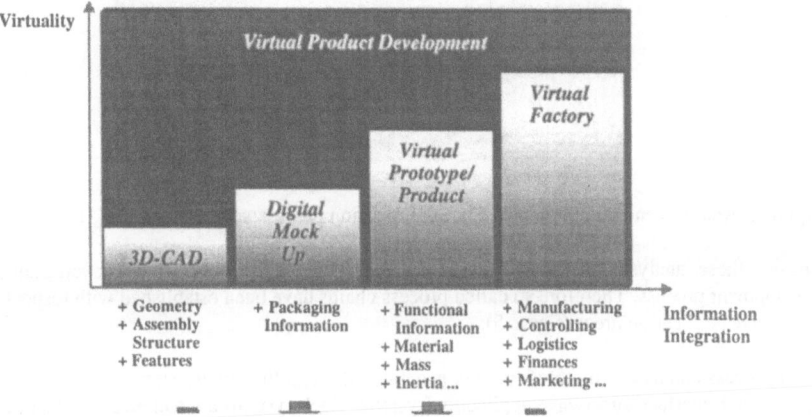

Figure 3: Overview of application software systems for Virtual Product Development

CAD-systems are often interfaced with Digital Mock Up systems (DMU-systems). They use the product structure and an approximation of the geometric representation and enhance these data sets by kinematic data. This enables an analysis of the kinematic behaviour of the product as well as the analysis of assembling and disassembling processes. Physical product data typically are not covered by DMU-systems.

Virtual prototypes contain physical product data with respect to their application specific domain, such as mechanics, electrics, electronics and also logistics (e.g. represented through software). Virtual prototypes typically are using Finite Element Analysis systems(FEA-systems) and Multi Body Simulation systems (MBS-systems). These systems are used for the analysis and the simulation of the physical product behaviour.

The Virtual Product is a digital product model covering the data sets generated through the application of the CAD-system, the DMU-system and the Virtual Prototype system. The development of Virtual Product analyses and simulation systems are of interest in research. Such a research initiative has been established as the iViP project (integrated Virtual Product Development [4]), where virtual reality is used for simulating the products behaviour.

The Virtual Factory uses the principles of the Virtual Product to analyse and to simulate the manufacturing, assembling and testing processes. Such approaches are e.g. developed for the virtual machine tool and the virtual manufacturing process. Figure 4 shows an application of a virtual manufacturing process analysing and simulating the part deformation process [5].

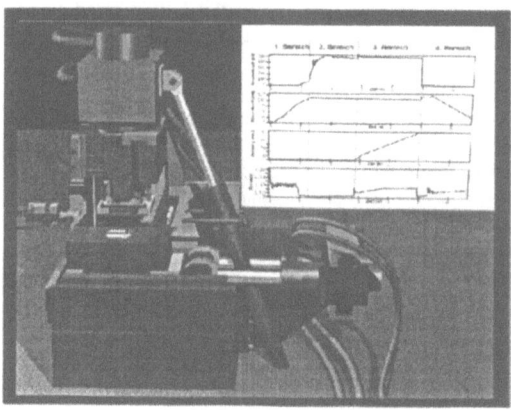

Figure 4: Analysis and simulation of a part deformation process;source Hemyari [5]

Besides these analysis and simulation tasks other objectives have to be covered in a product development process. Therefore so called process chains have been established with respect to support successive application areas (figure 5):

Such process chains use the product data generated through the engineering, development and design process and interface successive application systems. Such systems are technical product development (TPD-systems), computer aided planning and NC-programming (CAP- and NC-systems) and enterprise resource planning (ERP-systems). Based on a study from CADCIRCLE /CADCIRCLE 1998) the most applied process chain (56% of the applications) interfaced CAD and TPD.

As Virtual Product Development is not restricted to only one discipline such as mechanics, more disciplines have to be integrated into the product model, the multidisciplinary product model.

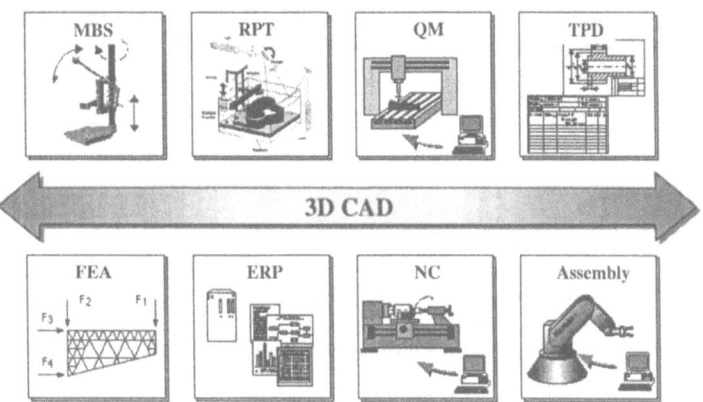

Figure 5: Process chains in Virtual Product development

3 Multidisciplinary Product Model

Products typically consist of a functionality that is based on various physical principles. Therefore such products are called multidisciplinary products. Most products today are multidisciplinary products. Due to the increasing importance of the combination of mechanics and electronics the name Mechatronics has been created. Mechatronics contains i.e. mechanics, hydraulics, pneumatics, electrics, electronics and even software for the control of its logical behaviour.

For the integrated analyses and simulation of the behaviour of multidisciplinary products it is necessary to develop an integration approach which is enables analysis and simulation systems to perform a multidisciplinary application and to communicate and to cooperate. Such an integration approach has been developed by the research projects EUMechatronik [6] and MechaSTEP [7][8]. EUMechatronik develops an integrated computer based engineering environment for the development and design of mechatronic products while MechaSTEP develops the integration platform.

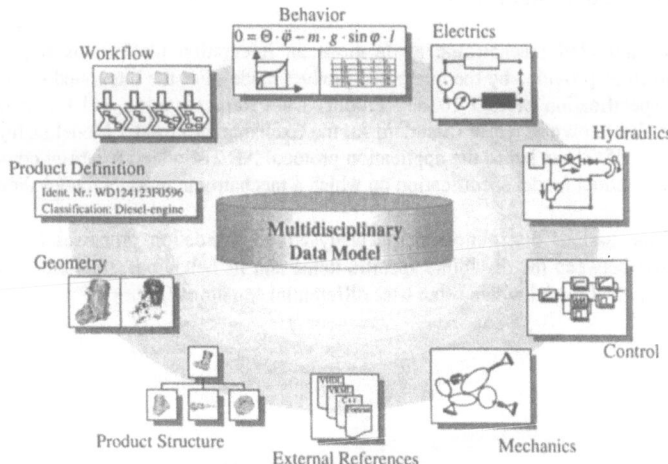

Figure 6: Modular structure of simulation modules

For the integrated computer based engineering environment an architecture has been designed which consists of a modular structure, containing various simulation modules (figure 6). This modular structure has to fulfil mainly two requirements. The reuse of modules and the non-redundant multiple application of the modules within a product.

This modular structure serves as a basis for the design of the architecture of the integrated computer based engineering environment (figure 7). The architecture shows four layers,

- the application layer with the various analysis and simulation tools,
- the system integration layer which controls the analysis and simulation tools,
- the process control layer where the workflow has been described, input and output information and process states are specified and
- the product data management layer where the mechatronic data model is managed.

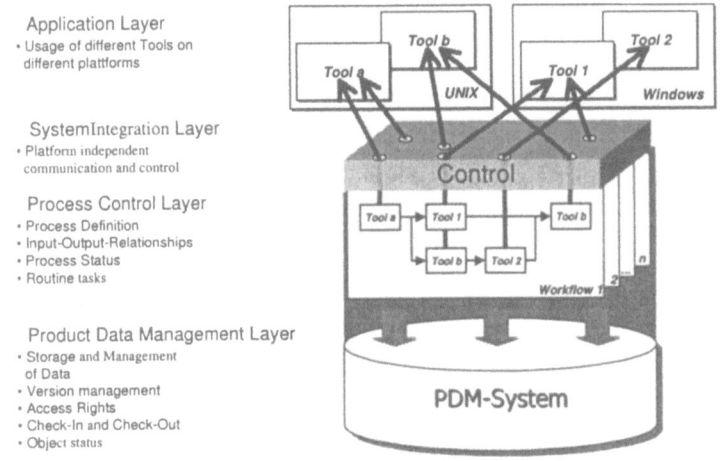

Figure 7: Architecture of EUMechatronik

For performing integrated mechatronic applications an integration platform is required. Such an integration platform is provided by the integrated product model. For the integrated product model the product model specification of ISO 10303 "Product Data Representation and Exchange" has been chosen which is also known as STEP (Standard for the Exchange of Product Model Data). An analysis of the STEP series /ISO/ has led to the application protocol AP 214 which has been considered as the most appropriate product model specification on which a mechatronic product model should be based.

An analysis of the various discipline specific analysis and simulation processes has shown that a dependency exists between the discipline specific items and its behaviour, the latter being described through a mathematical model which often uses differential equations (figure 8).

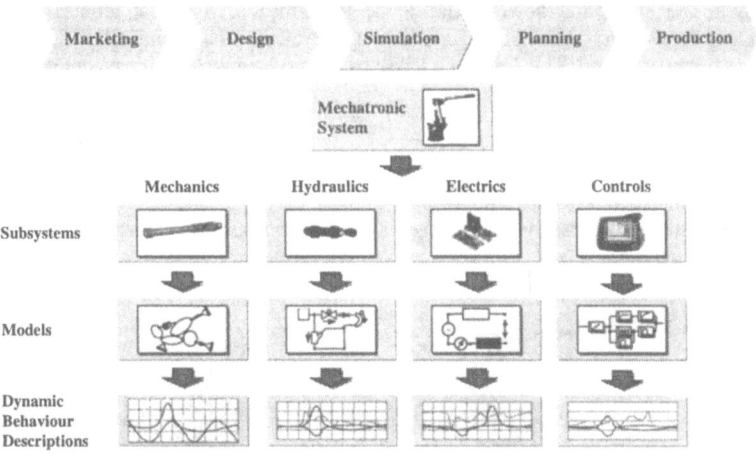

Figure 8: The analysis and simulation process of various mechatronic disciplines

As a result of the analysis of the various disciplines the object oriented data modelling technique for the specification of a mechatronic product model was considered as appropriate. A basic approach for the mechatronic product model is the representation of differential equations in the model. Such a representation has been developed starting with the mapping of differential equations onto a general graph and representing this general graph as an object oriented data model schema. This schema representation is then instantiated due to the mechatronic application. Figure 9 illustrates the mapping of differential equations onto an object oriented data model schema.

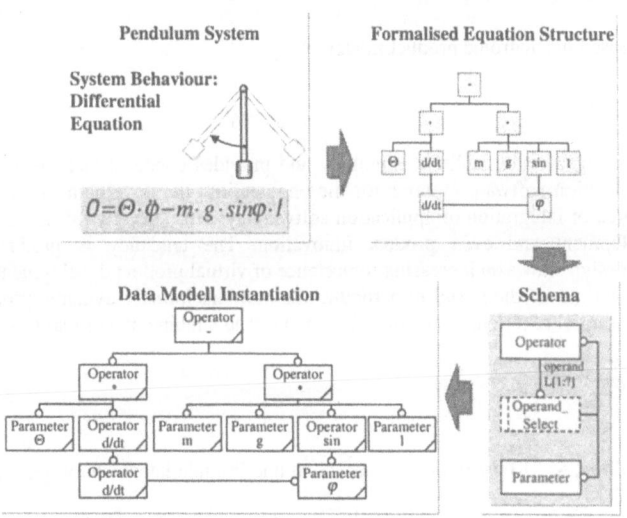

Figure 9: Mapping of differential equations onto an object oriented data model schema

From this object oriented data model schema object classes are being generated to created software for the analysis and the simulation of the behaviour of the mechatronic system.

The analysis and simulation however has to be performed as a virtual product. Therefore an integration of the dynamic behaviour description into the STEP product model and in particular into AP 214 has been performed. This leads to a STEP based mechatronic product model which presented by figure 10.

The STEP based mechatronic product model serves as an integrated product model description for implementing software systems. The implementation purposes cover communication between the analysis and simulation systems with respect to data exchange and data sharing. Besides data exchange and data sharing also system redesign is an issue and has been identified as being of interest for the software industry.

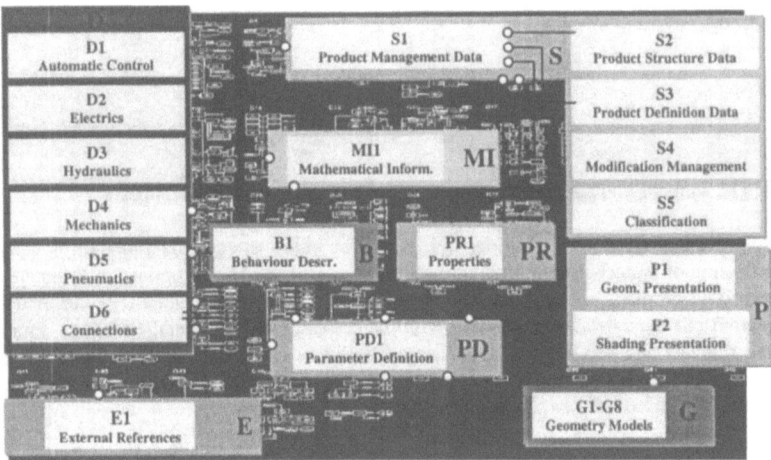

Figure 10: STEP based mechatronic product model

4 Conclusion

Product data technology is an enabling technology and provides concepts and specifications for the integration of application software systems for the engineering, the development and the design of products. The degree of integration of application software systems strongly influences the efficiency of product development and even product innovation. The tendency in product engineering, development and design shows an increasing importance of virtual product development methods. The integration of the various methods and in particular the multidisciplinary dynamic product behaviour description is an essential requirement to engineer, analyse and simulate the virtual product.

5 References

[1] Anderl, R.; Philipp, M.: Konstruktionswissenschaft und Produktdatentechnologie. Konstruktion, Heft 3/99, S.20-24
[2] Vajna, S.; Weber, Ch.: Informationsverarbeitung in der Konstruktion. VDI-Z, 1/99, S.20-23
[3] Spur, G.; Krause F.-L.: Das virtuelle Produkt: Management der CAD-Technik, Hanser Verlag, 1997
[4] Krause, F.L.; Tang, T.; Ahle, U.: Integrated Product Creation: Project Description IPK Fraunhofer Institute Production Systems and Design Technology, 1999

[5] Hemyari, D.: Methode zur Ermittlung von Konstitutivmodellen für Reibvorgänge in der Massivumformung bei erhöhten Temperaturen. Dissertation TU Darmstadt, 1999

[6] Anderl, R.; Claassen, E.; Krastel, M.: Produktdatentechnologie – Auf dem Weg zur integrierten mechatronischen Entwicklungsplattform..Thema FORSCHUNG Heft 1/2000

[7] Anderl, R.; Krastel, M.: Multidisciplinary Product Data Management (MPDM). Proceedings CME Conference, Bremen, 1999

[8] Anderl, R.; Donges, Ch.; Krastel, M.: MechaSTEP – STEP Datenmodelle zur Abbildung mechatronischer Systeme. Produktdatenjournal Nr. 5/1999, Darmstadt, 1999, S. 30-34

[9] Reinhart, G.; Grunwald, S.; Rick, F.: Virtuelle Produktion – Technologie für die Zukunft. VDI-Z, Oktober 1999, S 26-29

Get it Right the First Time

Collaboration in Heterogeneous Environments

Peter H. Ernst
OneSpace Research & Development - CoCreate GmbH, Sindelfingen
Peter-H_Ernst@bbn.hp.com

Abstract: Collaborative product development, long neglected due to the development of computer-aided tools targeted solely at individuals, has gained momentum with the availability of the Internet and the World Wide Web. Advanced computer aided product development environments put information at the fingertips of all people involved in the design, development, marketing, deployment, and purchasing of a product. Easy access to information and skills independent of geographic locations at any stage of the development process break down traditional barriers and allow designs with a quality and complexity not possible before. Although computer aided collaborative product development is more an information management issue, than a computer aided design and manufacturing issue, it is easiest understood in this very context, as here the development of computer-based tools have their biggest impact through computer graphics, 3D modeling and more recently virtual reality and digital mock up (DMU).

1. Introduction

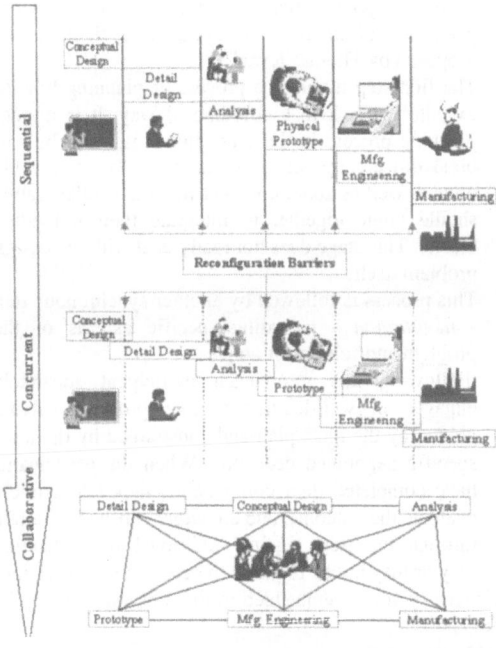

Very few people developing products today work alone because of the scale of the projects they are involved in and because of the range of disciplines required to come to a truly innovative solution. Thus, collaborative product development has been carried out with and without computer aids. The increasing availability and use of the Internet and the World Wide Web has opened up the opportunity for experts from different backgrounds to collaborate with each other in ways not possible before.

The impact of geography has changed dramatically. Customers, design and manufacturing specialists can now work on the same project at different geographic locations either synchronously or asynchronously.

Accordingly product development has itself changed from sequential to concurrent and, with the support of new computer technology, is now taking the next evolutionary step[2] to collaborative, interdisciplinary product development (see Figure 1).

The collaborative development process is best described by the connectivity of the various activities required to create a product, rather than by a chronological sequence of steps.

Depending on the problem at hand, particular activities are re-iterated many times in varying combinations. Collaborative computer tools actively support the initiation and fast execution of these *micro-cycles*.

 Cycles in sequential models usually indicate mistakes in preceding process stages, or even a not anticipated change in the product specification. The execution of cycles is hindered by *re-configuration barriers* (see Figure 1), making cycles awkward, costly, and time consuming. Re-configuration barriers prevent required changes and lead to sub-optimal products. Some common barriers are:

- Organizational barriers
- Geographical barriers
- Communication barriers
- Lack of Knowledge and Information

Collaborative tools and processes will do away with many traditional re-configuration barriers, but at the same time demand a fundamental change in behavior of individuals, solutions to overcome differences in time, culture, and work practice[4]. Experiences with remote collaboration demonstrate that social and organizational aspects overwhelm issues of technology[5]. Despite the tremendous success of collaboration tools for example in the area of Digital-Mock-Up (DMU) and computer graphics[1], these issues provide a yet to be mastered challenge.

In the following chapters, however, we will, adopt a technology-centric view and concentrate on the components to drive a computer aided collaborative product development in general and then move on to the special case of collaborative CAD.

2. Collaborative Product Development

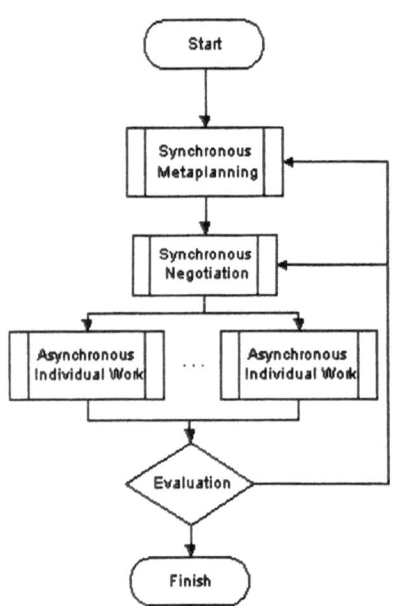

Figure 2. A High Level Model of Collaborative Product Development

Understanding the requirements for a collaborative product development environment requires some basic familiarity of the phenomenon of collaboration itself. In a general model of collaboration a project is composed of many micro-cycles (see figure 2) as proposed by Thomas Kvan[6].

The first step involves a process of planning how to execute the task in a coordinated way. It is a *meta* planning process in the sense that it is about how to break down the problem into individually manageable units as well as about how and when the collaborators should come together to integrate their individual efforts. This stage does not really deal with the design problem itself.

This process is followed by another synchronous step - *negotiation* - regarding specific aspects of the problem and solution strategies.

 Following this step, each participant separately engages in well-learned routine problem solving guided by the meta-plan and constrained by the task-specific negotiated decisions. When the participants have completed their components, they interactively evaluate the outcome, and are either finished or iterate through the steps again. Additional meta-planning may or may not be required to begin another iteration.

Considering the project context in which these micro-cycles are embedded further observations can be made:

- Participants have different technological background.
- Participants are geographically dispersed

- The computer-based tool environment is heterogeneous, i.e. each geographic location, maybe even each participant has a different computing environment.

The expected and in some research projects also proven effects[3],[5] are:

- Innovative Solutions. The iterative re-representation of information creates, according to Han J. Jun[8], a phenomenon called *emergence*, where new patterns and solutions are found by repeatedly looking at design problems from different viewpoints.
- Fewer design errors through timely access to data.
- Shorter time-to-market.

This can be translated into system requirements addressing aspects of synchronous and asynchronous collaboration.

To address the general and asynchronous aspects of collaboration, the environment must provide:

- A knowledge-base for information and data on a process , task, manufacturing, design, and analysis level and also problem domain level. Together with the artifacts of the design process this establishes the *design context*[7]. In interdisciplinary teams continuous access to information in the design context is essential to individual and shared understanding of purpose, function, behavior, and structure of the product under development.
- Multiple views on the elements of the design contest [9].
- A consistent, intuitive interface to information and data.
- Access control.
- A communication infrastructure for dispersed teams.
- The ability to integrate with existing corporate IT-infrastructures (i.e. firewall).
- Use of standards wherever sensible to enable easy data exchange in heterogeneous environments.

Aspects of synchronous collaboration to be covered are:

- Audio and video conferencing facilities. These are primarily important in early project phases, where the social context is built between the participants.
- Synchronous views of the data in the design context. This may be 3d models, meta-data, process data, etc.
- Collaboration management. This term refers to the coordination of synchronous sessions as well as to the documentation of sessions.

In the following chapters, we will explore the realization of these requirements in the special context of *OneSpace*, a collaborative CAD environment commercially available from CoCreate Software GmbH.

3. OneSpace - A Collaborative CAD Environment

The OneSpace collaboration system is based on client/server architecture (see Figure 3). Client or server can be located anywhere in a global network.

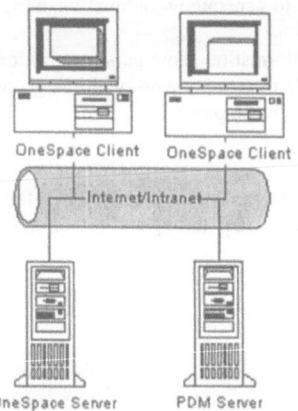

OneSpace Client   OneSpace Client

Internet/Intranet

OneSpace Server   PDM Server

Figure 3. OneSpace Architecture with Clients, Server and PDM

OneSpace Client

The OneSpace client is the front-end for integrated applications such as the 3D collaboration component or PDM interface. Clients can join a session (i.e. connect to a running OneSpace server) any time. Any client can trigger an upload of data into the OneSpace server for synchronous real-time viewing, inspection, conferencing, markup, and editing of the model. Results of the session can be downloaded to local discs, or stored in a PDM system. All upload and download requests, as well as all operations on the 3D models are under complete control of the configurable access control system. The lightweight Java client implements a consistent graphical user interface across different platforms. It uses a high-performance graphics subsystem based on the DirectModel framework for large-assembly visualization developed by Hewlett-Packard and EAI. The graphics subsystem achieves high rendering speed through frustum culling, occlusion culling and different levels of detail. While the geometric model resides only on the modeling kernel of the OneSpace server, its graphical representation, structural information, and meta-data is sent to the clients. No intermediate files of data format is involved. Server and client talk directly to each other. During the streaming process, the receiving clients render the 3D model part-by-part as they are received from the network. All relevant attributes for shaded rendering are received together with the tessellated model. The local data set also contains the identification of all model entities such as assemblies, parts, faces, edges, vertices so that the client is able to identify elements of the 3d model when sending requests to the server.

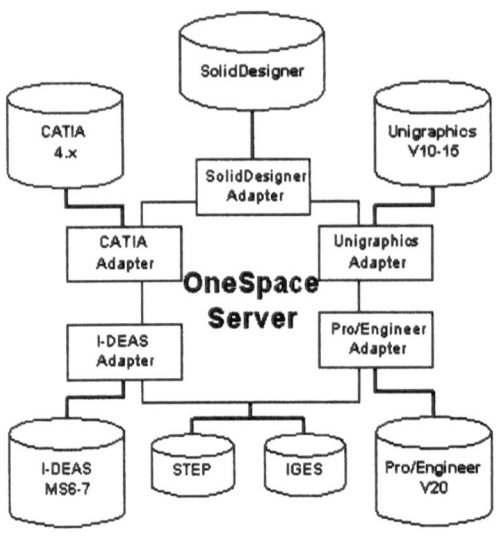

Figure 4 Direct CAD system links

OneSpace Server

The OneSpace Server includes a complete solid modeling kernel that holds the 3D product model. The kernel architecture is optimized for data exchange, and interoperability with multiple CAD systems.

Direct Links to CAD Systems

In the OneSpace solution context, the quality and information content in heterogeneous CAD environments play a central role. Beyond the standard file exchange formats (STEP, IGES), also direct links to other CAD systems have been developed. These direct links allow optimized transfer of 3D shape data as well as associated information. Associated information plays a critical role in the collaboration context because it conveys a broad spectrum of non-geometric information essential to understanding purpose, function, behavior, and structure of the model.

To optimize the quality of 3D shape transfer CAD data files can be accessed directly through adapters, which are optimized for the particular model representation of the connected CAD system (see figure 4). This optimization is bi-directional. When reading native files, the corresponding CAD adapter use special knowledge about the originating system to generate an optimal OneSpace model. When writing native files this special knowledge is used again to generate an optimal model for the receiving system.

If the data comes from a parametric, or history-based system, information about parameterization and feature structure is lost in the import process. In order to be able to actually work with the geometric model non-parametric and non-historic modeling algorithms must be employed.

Modeling

The OneSpace Server uses a boundary representation (B-rep) of the model. An architecture referred to as *Dynamic Modeling*, provides the ability to manipulate the model through a set of advanced operations without the knowledge of how the model was created:

- Feature recognition
- Model deformations
- Rounding operations
- Cutting and pasting of face-sets.

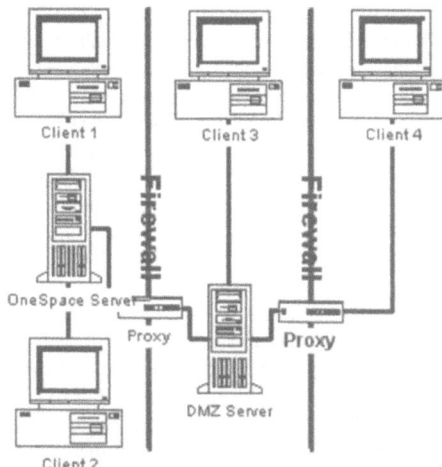

Figure 5. Cross-Firewall Communication

Incremental Client Updates

If a new client connects to a session, the server sends an individual copy of the graphical model to the new client. This ensures that every participant is working with the same data in its most recent form. If models have changed in response to a client request, only incremental updates are broadcast to all connected clients. Once a model has been loaded, the subsequent incremental client updates consume only a very small bandwidth on the network connection.

Client/Server Communication

The OneSpace framework operates in a standard TCP/IP local area (LAN) or wide area (WAN) infrastructure. This has the following benefits:

- It is easy to integrate in existing corporate networks.
- It works with dial-up connections.
- It works over the Internet.

Crossing the Firewall

The OneSpace architecture also provides a *Firewall* communication (see figure 5) using *HTTP tunneling*. HTTP tunneling is a method to encode data in the HTTP protocol and sent it as HTTP request, receiving HTTP encoded data in return. This method allows client/server communication across the firewall using HTTP proxy servers as gateways.
Because of the filtering nature of the HTTP proxies there needs to be an additional server in a *demilitarized zone* (DMZ). The DMZ-server acts as a message relay.

Collaboration Functionality

In this chapter, the implementation of collaborative functionality based on the previous architecture is outlined. From the users perspective collaboration happens on the client side. This is the piece of software she/he interacts with. Consequently, the description of the collaboration functionality mainly refers to the OneSpace client.

Viewing and Inspection

Once the initial load phase is complete, each client is equipped with all data needed to render the model locally and independently of other clients and of the server. This independent viewing mode is very important for participants to get familiar with the model. A participant can also select elements of the model on the local display. Since clients also receive structural information relating to assemblies, parts, faces, edges, vertices, etc. there is an association to the exact model on the server. Hence, a client can address model-based requests to the server to make precise measurements of volumes, areas, lengths, clearances, etc.

Conferencing

The key need for unambiguous communication is the ability to:
- See the model from the same perspective
- Mark areas of the model
- Share documents
- Use a shared whiteboard

- Talk to each other

With the OneSpace clients participants can share views with selected partners. Since all clients have complete local rendering information only viewing parameters need to be exchanged. These parameters include camera position, viewing direction, and zoom factor. Even very low bandwidth connections, like through a 28.8kbit modem, is sufficient to achieve real-time synchronization.

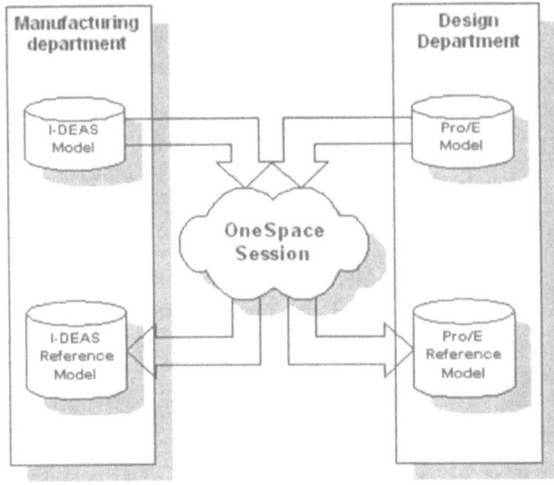

Figure 6 A Collaboration Use Case

At times, where the synchronization of views is not sufficient, model markup can be used to achieve unambiguous communication. The available markup methods are:

- Shared pointer: A 3-dimensional arrow which can be dragged across the surface of the geometric model. It is visible by all participants, but is volatile. It does only exist in the context of a OneSpace session.
- Notes: Textual information or hyperlinks (URLs) can be attached to geometric elements. This information becomes part of the model and therefore is persistent. The native CAD adapters will transfer notes to their associated CAD systems,

where they can be used for reference.

If looking at 3D data is not enough, views on any data residing on any participant's system can be shared with others by the integrated application-sharing module.

Upon request, the integrated shared whiteboard can be opened and any notes or sketches can be collaboratively developed.

A voice/video connection can be established using the integrated conferencing component. However, in most cases a parallel phone conference via regular phone lines is more appropriate.

Collaborative Modification

Viewing, conferencing, and inspection provide the ideal foundation for collaboration to identify problems wit the virtual product. However, problem resolution requires a further step. Since a solid modeling kernel is part of the OneSpace server, modeling requests can be sent to the server to perform model changes based on the preceding discussions. Any model change is automatically logged including all parameters and details of the change. Again this information is persistent and will be transferred to the CAD system along with the geometric shape. This data enables the designer on the receiving CAD system to follow up on the collaboration session with accurate and comprehensive information.

A Typical Use Case

To further explore the application of computer-aided collaboration a typical scenario involving the communication between a manufacturing department and a design department at different geographic locations is described. A sketch of the process flow is illustrated in figure 6.

The manufacturing specialist is not only responsible for the assembling the various components and the manufacturability of the product; he must also keep process cost to a minimum. He uses I-DEAS for analysis and simulation, where as the design department uses Pro/Engineer for parametric design.

The design specialist is responsible for aspects of function and form on a component of the product. He uses Pro/Engineer for parametric design.

The design and manufacturing specialists must get together at regular intervals to check issues of manufacturing and assembly.

1) In preparation of a collaboration session, e-mail is sent to both specialists. It contains includes: Meeting date
- A server address
- A session password
- A Phone number and ID for a phone conference

2) Before the scheduled meeting time the manufacturing specialists starts his OneSpace client and:
- Configures the collaboration server using the server address from the e-mail.
- Logs into the collaboration session using the session password from the e-mail.
- Pre-loads the I-DEAS product assembly into the running session. In preparation of the online session, he may have added product manufacturing information in the context of his CAD system.

3) At the scheduled meeting, time the both specialists dial into the phone conference. Additionally the design specialist starts his OneSpace client and connects it to the already running session using the configuration and login information from the e-mail.

4) As soon as the design specialist is online, his name appears in the *members window* of each client next to the manufacturing specialists name.

5) The design specialists uploads the Pro/Engineer part, he has been working on for this product.

6) Now the manufacturing specialist moves the uploaded part into its correct position using the appropriate positioning commands of the client.

7) Now, both specialists discuss the uploaded part using independent views, shared views and 3-dimensional pointers. They inspect geometric properties, interferences, and clearances as needed.

8) Associated information (notes) is attached to elements of the uploaded parts in order to:
- Document decisions
- Record ideas, or remarks
- Hyperlink to relevant technical documents somewhere on the web

9) Local documents are shared through the integrated application-sharing component, and sketches are drawn on the shared whiteboard.

10) The uploaded part or the pre-loaded assembly is modified in order to explore design optimizations, or alternatives. Changes are performed independent of the parameterization and feature structure of the original model. Each model change is automatically documented by the system.

11) When all participants are satisfied, the resulting models, decorated with notes, and URLs are downloaded to the local computer systems. Later they are imported into the corresponding CAD systems to follow-up on the online session. As the imported models lack the parameterizations and feature structure of their corresponding originals, they are used for reference only. The attached information helps to make the indented changes on the original part the context of its proper parameterization and feature structure.

To keep things simple the role of PDM has been neglected in the use case above. In most cases a PDM system will be online during the collaboration session. PDM access is an integrated part of the OneSpace client.

Use Case Summary
- OneSpace uses a *sandbox* approach, where changes can be safely explored without affecting the original models
- Parts from different CAD systems can be assembled in one session.
- Textual information and hyperlinks can be associated with shape elements.

4. Summary

Heterogeneous CAD environments have been used to analyze the contribution of computer-based collaboration to product development. The most important contributions were found to be:

- Import of CAD models from different sources into one model-space for inspection and change regardless of their history or parameterization.
- Unambiguous communication through high-fidelity computer graphics and collaboration tools.
- Decisions can be made *right on the spot* through availability of shape information as well as related data (meta-data).
- Feedback across geographical or structural department boundaries can be obtained at any stage of the product development process.
- Documentation of collaboration sessions.

5. Legal Notices

- I-DEAS™ is a registered trademark of Structural Dynamics Research Corporation
- CATIA™ is a registered trademark of Dassault Systems
- OneSpace™ is a registered trademark of CoCreate Software GmbH
- Pro/Engineer™ is a registered trademark of Parametrics Technology Inc.

6. References

[1] Kent Beck: *Embracing Change with Extreme Programming,* IEEE, 1999, IEEE Computer Journal
[2] Arnold Müller: Shared Engineering with OneSpace, 1999 Technical Whitepaper, CoCreate
[3] Robert M Abarbanel: *Flythru the Boeing 777,* 1997, Formal Aspects of Collaborative CAD, pp 3-9
[4] Les Herbert, *Collaborative Design on Global Projects,* 1997, Formal Aspects of Collaborative CAD, pp 11-14
[5] Dean Taylor, Kevin O'Connor: *Experiences with Remote Collaboration for Concurrent Engineering,* 1997, Formal Aspects of Collaborative CAD, pp 29-47
[6] Thomas Kvan, Robert West, Alonso Vera: *Tools for a virtual Design Community,* 1997, Formal Aspects of Collaborative CAD, pp 109-123
[7] Ming Xi Tang: *An Architecture for Design Collaboration Management,* 1997, Formal Aspects of Collaborative CAD, pp 217 - 236
[8] Han J. Jun, John S Gero: *Representation, Re-Representation and Emergence in Collaborative Computer-Aided Design,* 1997, Formal Aspects of Collaborative CAD, pp 303-319
[9] Michael A. Rosenman, John S. Gero: Collaborative CAD Modeling in Multidisciplinary Design Domains, 1997, Formal Aspects of Collaborative CAD, pp 387-403

Living Data as Knowledge Source

H. Grabowski, R. Ostermayer

Institute of Applied Computer Science in Mechanical Engineering (RPK), University of Karlsruhe, osterma@rpk.mach.uni-karlsruhe.de

Abstract: Living data are data that are in use or that are (re-) useable for human beings and machines. In other words, living data are information with a distinct semantics in a considered context, that are valid within a distinct period, that are available and accessible, and hence, that are (re-) useable for human beings and machines. In this paper three examples will be explained how to keep data alive and how to use data as knowledge source. The examples to be discussed are the automatic classification of products, a function based information retrieval system, and as a consequence a semantic framework to manage complex information models.

1 Introduction

The current situation on the market is signed with activities of rationalisation to a big extent. Reasons are the increasing pressure of competition and contest for the branches due to the globalised market and the change of the economic systems. The demand of the market towards more powerful products results on one hand side in an increasing variety and on the other hand side towards an increasing complexity of products. The increasing product functionality and complexity, shortening of the time to market with increasing quality, and reducing costs put more and more requirements to the designers and to the structure of the process of the development as such.

As a result of this evolution, more and more information and knowledge is required for the task at hand. Especially the early phases of the product development, i.e. planning and conceptual or preliminary design, are signed with huge information and communication flows related to the complete product development and production process. For instance, the indirect design tasks, such as collection of information, knowledge acquisition, or talks with clients, are about 40 % of the design tasks [1], [34]. While passing times in the production are permanently decreasing by means of rationalising processes and introducing new technologies, in the area of the early phases there are huge backlogs with respect to methods and techniques for building efficient information and knowledge management structures and systems. However, the information or knowledge acquisition plays an important role especially regarding a fast and cheap product development: Decisions for the future are made in the beginning of the product development but not at the end during production or assembly, respectively [13].

In this paper, we want to show some aspects of rationalisation in terms of organising the data, information, and knowledge gained during product development and the re-use of this knowledge for a current problem. The presented systems are for the automatic classification of products, a tool for the effective information retrieval, and a semantic framework for the representation of information and knowledge.

2 Current Approaches and Problems

The theme of the workshop is "Living Data is the Future"! However what is considered as living data or information, respectively (*living* data implies a higher semantic level than just data hence we use the concept of *living data* and *information* synonymy) and how can we keep data alive.

If we regard the development process of a car, we can distinguish different stages and aspects starting with the product definition and ending with recycling. Each of the stages creates and/or re-uses data and information. For instance, the requirement specification within the stage of the product definition, has impact onto the complete product lifecycle, i.e. if the requirement *manufacturable in-house* is hurt, it could come to a delay to put the product on the market which in turn is coupled with losses of money. Figure 1 shows the creation and re-use of data in the automotive domain. In this figure only the impact from the perspective of product definition is shown. The interdependencies of the different sources are like a network which connects each stage to each other.

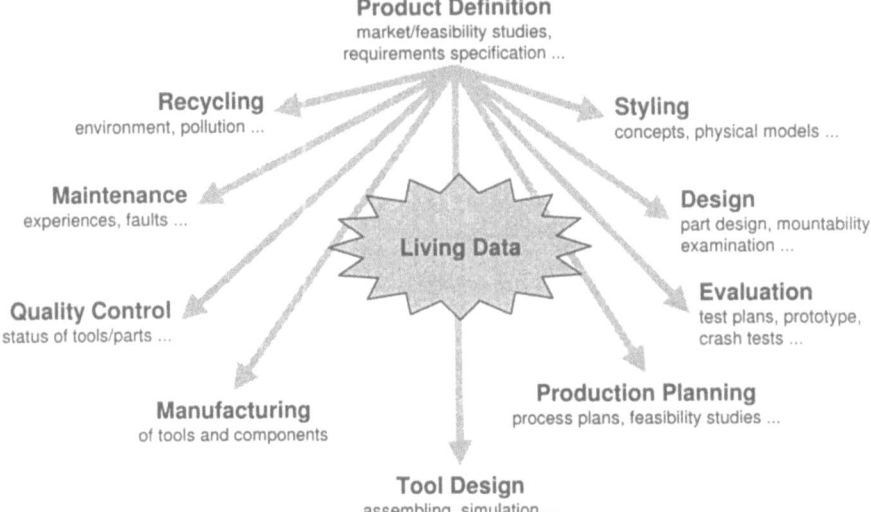

Figure 1:   *Creation and Use of Data (e.g., automotive domain from the perspective of the product definition)*

Summing up, living data are data that are in use or that are (re-) useable for human beings and machines. In other words, living data are information with a distinct semantics in a considered context, that are valid within a distinct period, that are available and accessible, and hence, that are (re-) useable for human beings and machines.

Regarding current CA-systems (i.e. Computer Aided systems, such as CAD, CAE, CAM, etc. for design, engineering, manufacturing, etc.) and consider the above mentioned requirements, two main directions may be distinguished, i.e. data representation and exchange, and the documentation, presentation, and management of data.

Data representation, i.e. the internal representation model of the underlying system which represents the structure according to (syntactical) form and (semantic) contents in order to represent the intention and semantics of the appropriate model. Contents is according the definition of object classes and their relations among each other. Form is determined by the data types of the attributes. Integrated product and production models contain information from different viewpoints, such as design and production planning. For instance, a design model is created and is then passed to the production planning for creation of a digital mock-up in order to examine assembling of components with robots or to do strength examinations in terms of a digital crash test when evaluating the design. Depending on the degree of integration, an integrated product and production model enables the (re-) use of the created instances in different production stages.

Data exchange among different systems, such as between a CAD-system and a PDM-system (Product Data Management), is done in the native representation formats of the CA-systems, with standardised interfaces, such as IGES (Initial Graphics Exchange Specification) [2], VDA-FS (interface specification for surfaces from Association of German Car-Manufacturer VDA) [11], STEP (ISO

10303) [37], or with relational and/or object oriented databases (such as Structured Query Language, SQL [4] or Common Object Request Broker Architecture, CORBA [25]).

Data documentation, presentation, and management is done with PDM-Systems on the basis of an integrated product and production model [14], [7]. Those kind of systems gather all product defining data together. Up to which extent all single entities are accessible depends on the internal representation model.

However, how is the context represented in which the data and information have to be seen. Context is represented implicit but context is the prerequisite for interpretation and (re-) use. Every kind of data becomes information if the semantics and the context is known. Different comprehension or interpretation of the data leads to ambiguity and uncertainty which in turn is a big problem for the (re-) usability by humans and machines. Furthermore, knowledge represents the structure, meaning, usage, rational, interpretation, and further functional properties of data [31].

Most of the systems, however, are concentrated on documentation, presentation, and management of the data but not how the problem was solved. One exception is the model of SFB-346 (a research project at the University of Karlsruhe) [16]: here also design-process-pattern and design patterns are considered to capture experience and knowledge gained through previous tasks, problems, and solutions.
An other exception is Cyc^{TM}, i.e. representation and reasoning about common sense knowledge. Common sense knowledge has to take contexts into account since humans are always talking and doing things within a distinct context which is usually recognised by humans but not by computers. Cyc was initiated as a research project in 1984. Since 1995 Cyc is developed from Cycorp Inc. Objective of Cyc is to model knowledge which is used to understand common-sense-knowledge. This kind of common-sense-knowledge may then be (re-) used to investigate and interconnect databases or for natural language interfaces between humans and a computer [33].

However the reality in companies looks different. The big European enterprises loose billions of marks due to bad organised information networks. According to an estimation, it is about 50 billion $(50 * 10^9)$ Deutsche Mark since the information retrieval in the companies is too difficult an not effective. This are approximately 600.000.000 working hours a year for the search of information whereby the success to get the right information for decisions to be made is only 30% [10].

In terms of having the right information, at the right time, at the right place, with appropriate quality, a product classification system, an information retrieval system, and a semantic framework is presented in the following.

3 Classification, Information Retrieval, and further Considerations

In the following we want to show a classification system which automatically classifies product properties taken from an integrated product and production model. The classified product defining data enable the re-use of components designed in the past but also concepts of solutions which went never in the production.
An other system to be shown is an information retrieval system which considers the context of given problem which enables to express the problem and not the solution which should be the result of a query.
Finally, a semantic framework is proposed. The semantic framework is a consequence of the representation of information and knowledge in order to have one representation system for the knowledge but multiple systems which re-use the stored knowledge, and hence it is a brick for formal knowledge management.

3.1   Automatic Classification of Products

The process of classification is an approved method to order data, information, and knowledge units similar to the way of humans thinking. The process of building classes is leading to the creation of new concepts if similar objects are collected together due to their specific properties. Hereby, a reduction of the data is achieved [3], [8], [9]. If we regard the product development process and its increasing need of information, and in turn as a result, the enormous increase of the volume of information being created, the classification is an appropriate aid to pack data and information to a semantically higher level [24]. The product classification based on similar properties of items is a way to provide product knowledge in a compressed way for the application specific usage within all departments of an enterprise (Figure 2).

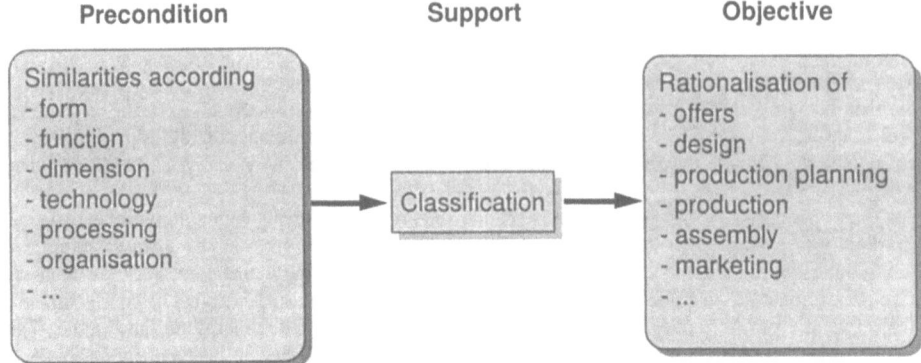

Figure 2:   *Preconditions and Objectives of Classification*

The effective access to stored data and the re-use of solutions for a new task at hand has nowadays a strategic meaning since the pressure of costs and time to market became the success factor number one [41], [39].

Examinations confirm that the number of parts may be reduced to 10% to 40% if an effective access to the product defining data is provided [24], [42]. On the other hand side, the introduction of a new DIN-A4 drawing including the derived product information causes costs of about 2,500 to 4,000 DM. The complete life-cycle costs are about 28,000 DM [32], [20].

First examinations about efficiency of product classification and analysis of similarities goes back to the 60s. According to Dirzus [12], the implementation of a classification systems with manually classification leads to 2% additional time for the classification process since 2 additional people per 1000 designers prepare the information for re-use. However, the design time frame after initiation of the product classification was reduced to 90,5% to 55% of the original time (Figure 3) since the re-use of former solutions led to the creation of part-families for production planning, manufacturing, etc. and hence, it decreased the number of variants and the product complexity.

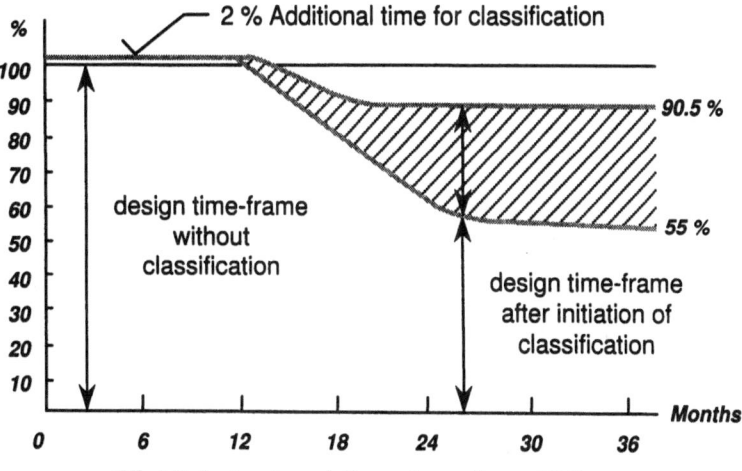

Figure 3:   *Effort Reduction through Re-use (according to [12])*

Those former methods had a lot of inefficiencies since a lot of work for the classification had to be done manually. The automatic classification of products based on the concepts of HAIN was further developed and implemented at RPK [24], [19], [18] and will be briefly explained in the following, an example classification derived from a daily problem will illustrate the system.

The concept for the automatic classification of products leads to a system architecture where the integrated product and production model (PPM) plays a central role. The PPM persistently stores all product defining data and thus it enables product describing information along the life cycle of a product. The basic idea of the system for automatic classification is that products are described with CAx-systems along the product development process and classification considers distinct, application specific views onto the stored data. The strategy of the automatic classification is to use the declarative knowledge of the represented product properties within an integrated PPM and to use appropriate methods to gain specific classification keys from the representation of those properties.

The system for the automatic classification is divided into three main modules: A general data interface, a complex data interface, and the classificator (Figure 4).

The general data interface enables the logical integration of data with an integrated PPM. Here different CAD-representation models may be mapped onto the PPM in a way that they are directly usable within the classificator. For instance: The geometry of a cylinder may be represented as a circle which is extruded to a distinct length, along the normal direction perpendicular to the circles plane.
The main application, however, is the re-use of these objects for the flexible definition of complex classification properties which is done within the complex data interface. Here, elementary product properties taken from the PPM are used to derive complex properties. This is done with mapping methods and calculation rules. Derived properties are user defined and they are used to create new classification classes and their hierarchies, and to create new classification schemes appropriate to the considered view. For instance: According to Opitz [35], [36] special lathe parts are discriminated with respect to the ratio between the length L and the diameter D. If L/D $\geq$ 3 it is considered as a long lathe part which may be in turn manufacturable on a specific machine.
Finally, the classificator does the examination of the user data based on the defined classification schemes and then presents the results in a form that is readable by humans.

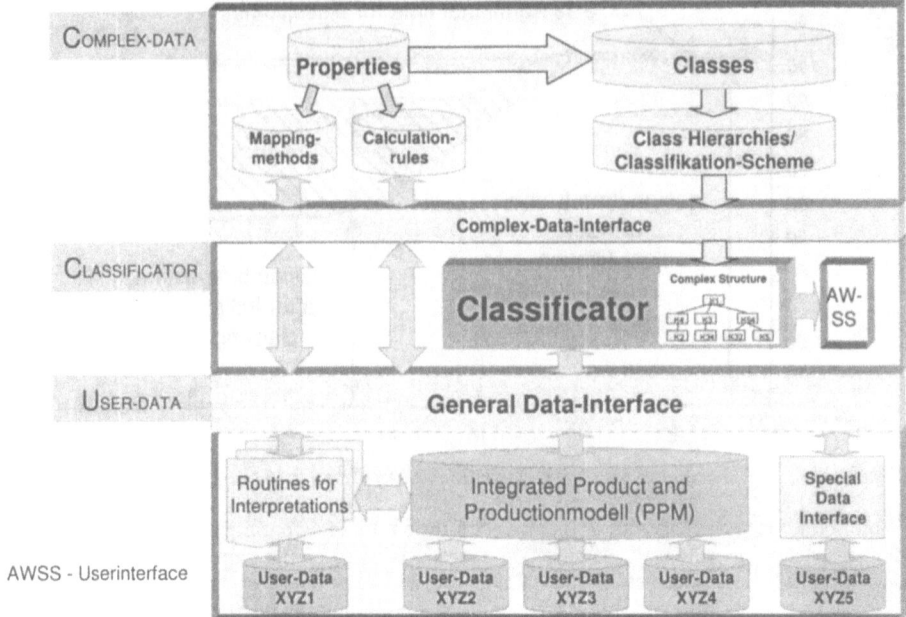

Figure 4:   *Architecture of the Classification-System*

The next figures demonstrate the application of the automatic product classification. In the example, the raised question is: which lathe parts can be manufactured in-house on which machine.

According to Figure 5 window (1), parts are distinguished according the production aspect, i.e. parts which are bought from a supplier or parts which should be manufactured in-house: possible or not is the question. If parts are to be manufactured in-house then the dimensions of the distinct part may not exceed dimensions given by the machines or production systems. In this sense, a rule has to be defined which takes different properties of the machine into account, such as maximum weight, size, etc. in order to prove the processing of the part as it is shown in window (4). The complex property is now denoted as "manufacturable on own machine" (see window 2). The structure of this property is shown in window (3).

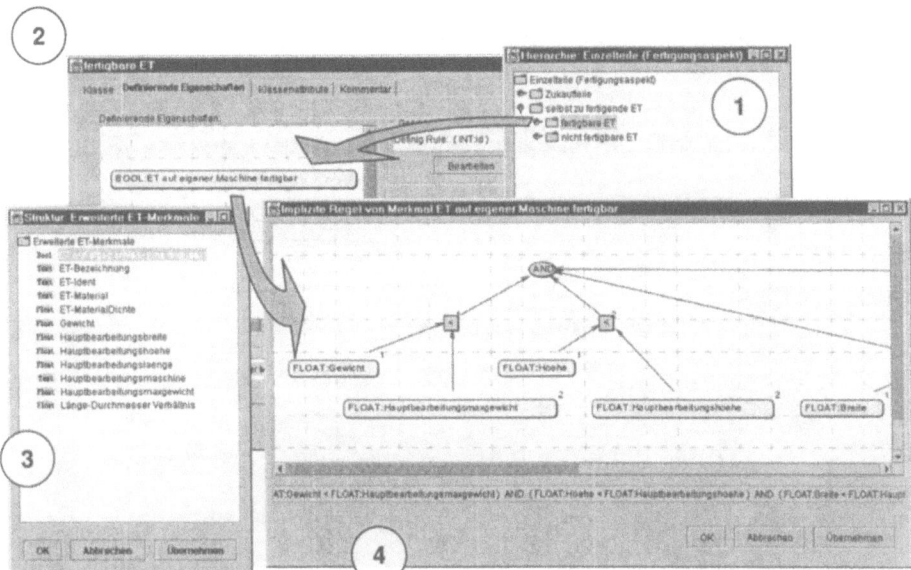

Figure 5:   *Rule definition as example*

Since we want to see the interconnection between processing machine and a part to be manufactured, we also need a specialised part. Here we consider a long lathe-part (remind definition above). The definition is shown in Figure 6.

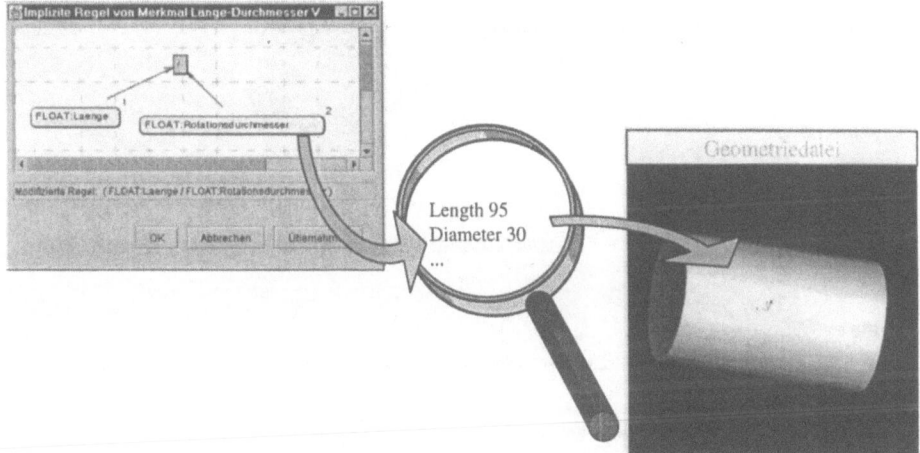

Figure 6:   *Definition of a Long Lathe Part (L/D ≥ 3)*

Figure 7 shows the relation between a part and the processing machine. In the Integrated Product and Production Model the relation between Part, Processing Machine is given. There is also a possibility to add a path between part and machine manually. The dependencies are graphically presented.

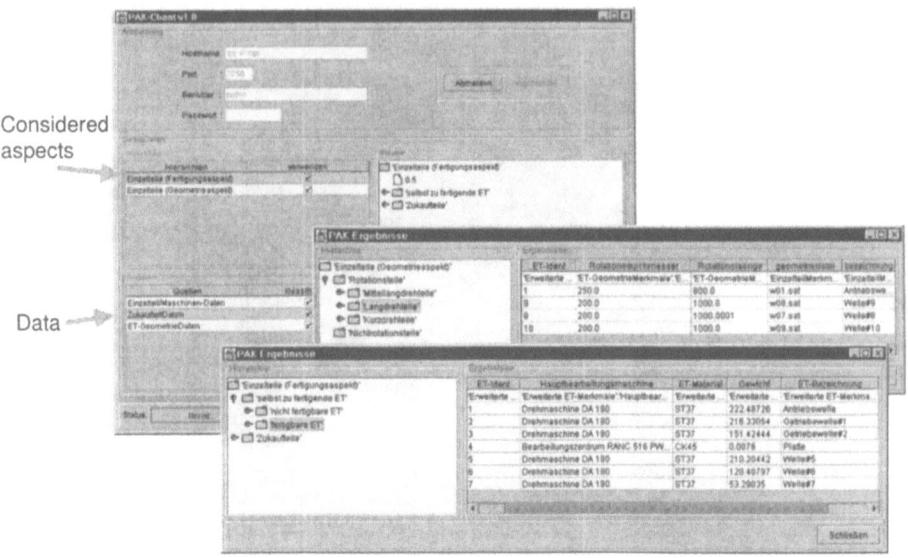

Figure 7:   *Path between Part and Processing Machine*

Result of the Classification in terms of processing machines and long lathe parts which are processed in-house is given in Figure 8. The table in the upper window is according to the structure of window (3) in Figure 5.

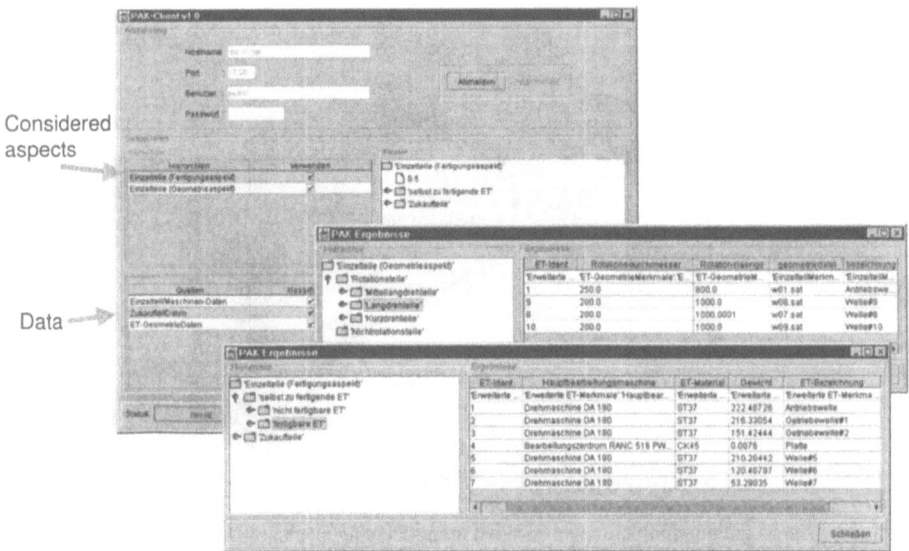

Figure 8:   *Result of the Classification*

The development of the automatic classification system is not finished yet. Further industrial requirements will fit the system towards more flexibility from the perspective of different domains.

3.2   Function Based Information Retrieval

Information retrieval considers the technical supported process of knowledge transfer from the knowledge producer to the knowledge requester. Information retrieval as a research area was developed in order to use the computer for administration of and search for textual information. Information retrieval functionality is especially interesting if vagueness and uncertainties are important aspects with respect to inquiries and the stored knowledge. The prototypical recherché situation differs in several aspects from the standard use of a database management system [27]:

- The semantic of the stored data is fuzzy and not complete with respect to the inquiries, since the information retrieval system is not operating with the contents of a document but it considers a document as a sequence of words.
- The inquiries are fuzzy, since often the inquiring person can only precise the information needs in dependence of intermediate searching results.
- The quality of the retrieval results depends on the selection of the proper database, the completeness and actuality of the database, the formal preparation of the database, the preparation regarding the contents of the database, the offered retrieval functions, and last but not least the skill and experience of the inquiring person.

Since initiation of this research area, retrieval systems provide only few methodological innovation [6], [26], [29], [28], [38], [34]: The search in simple text allows to search for every occurrence of a word form. Additionally string oriented operators are available for covering word variants: right and/or left truncation and masking of words. Context operators allow searching for multiple words with a distinct spacing in a sentence or database field. Special search functions are for instance: Basic word form recherché which finds words independent of their flexion (search for *mouse* also finds *mice*), decomposing of words in order to replace and to precise the left truncation, phonetical recherché which leads to equal or similar sounding words, and support of multilingual search in order to find keywords independent of the language.

In many cases current information retrieval systems deliver only unsatisfying searching results. Either the number of answers is not overlookable or actually relevant information are being filtered. The main problem is: classical retrieval systems do not operate with the contents but with a sequence of words. Unclear inquiries amplify this problem. However, as mentioned in the introduction, about 40% of the design time is for information acquisition [1], [34].

In order to avoid the problems of classical retrieval system, the function based information retrieval for the semantic processing was developed [17]. The idea for such a system is based on the function oriented view onto products and objects as it is done within the mechanical engineering domain. Here a function may be understood as an abstract described, general operating relationship between input, output, and state parameters of a system for the fulfilment of a task [39].

In other words, technical problems may be represented within a functional context. The context represents the relation between cause and effect of the objects to be investigated. User profiles about the considered domain and additional constraints on input and output objects help to limit the hit rate of a query. For instance: Find a machine element, or connection, respectively, which passes a torque. The technical problem is, how can a torque be passed? In the mechanical engineering domain a torque may be passed with a machine element, e.g. a key. The fully specified functional context is then passed to a document archive where the inquiry is done.

Such an abstract view, based on functional contexts, enables a problem specific enlargement of the space of solutions for inquiries since the retrieval of solutions is done within a given, distinct context, i.e. non-relevant information is excluded. The solution-space of a request is enhanced since the detection of more abstract solutions is possible and thus the generation of new potentially solutions is given. The concept of the function based information retrieval is shown in Figure 9.

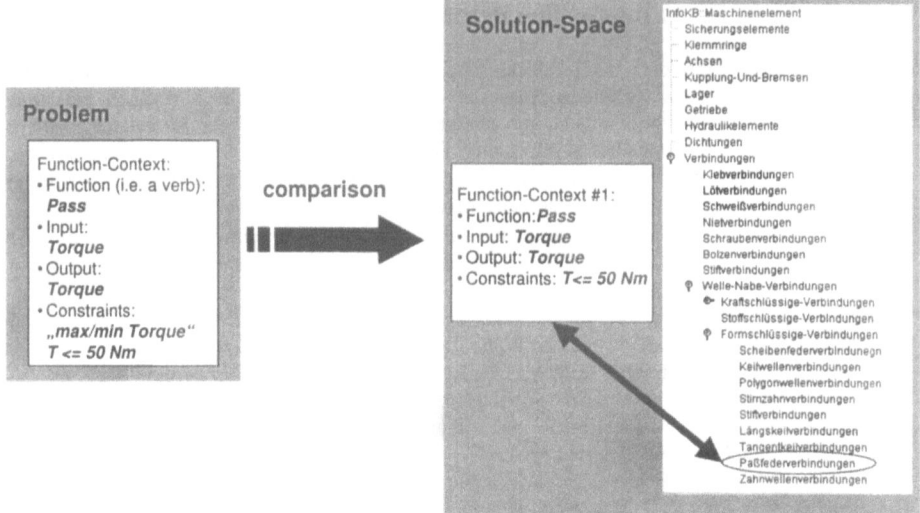

Figure 9:   *Concept of the Function Based Information Retrieval*

The software for the function based information retrieval is developed with Java 1.2 and Open Knowledge Base Connectivity (OKBC) [5]. It has a client-server structure and is able to run on the internet. The number of clients connected to the internet is not limited by the application. The architecture of the system is shown in Figure 10.

The system consists of four main modules, i.e., a client application, a method server, an archive server, and a knowledge server. The client provides functionality for the connection to the internet, for the search inquiry, and for the actual search. Through the API (i.e., application programming interface) the command with its parameters will be transferred from a client to the method server. The method server provides all implemented functionality required for the function based information retrieval. The archive server provides persistently stored documents and the derived index files. The knowledge server provides access to the knowledge bases where the functional contexts are stored.

The communication between client and server is done by the OKBC-protocol, sockets, and the World Wide Web.

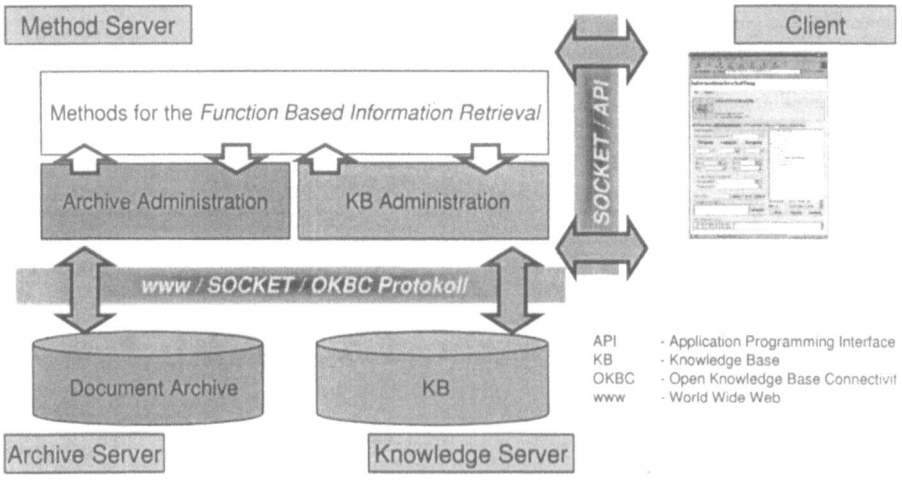

Figure 10:   *Architecture of the System for the Function based Information Retrieval*

The next figures demonstrate the application of the function based information retrieval with respect to the client connection and the wording of a search inquiry (the search process within documents and the incorporation of new knowledge and knowledge bases is not shown here since they do not provide a deeper understanding of the presented concept). Since one requirement was that the application is running on the internet, the user interface is very compact and the information within the window changes as required.

In the example, the problem to be solved is: Find a machine element which passes a small torque and which is simply mountable.

At very first, a client connection has to be established (Figure 11). Connections may be established to wherever a compatible knowledge base is stored. The user has to provide a user name and a password to get a connection. The type of the knowledge base may also be selected if different knowledge representation systems are used. If logged-on, a list of all available knowledge bases is shown in the in the lower right window. At the bottom, a window for system messages from the method server is displayed.

Connection of a
client-application
to a knowledge base
(KB) and log-on

Selection of
the KB-typ

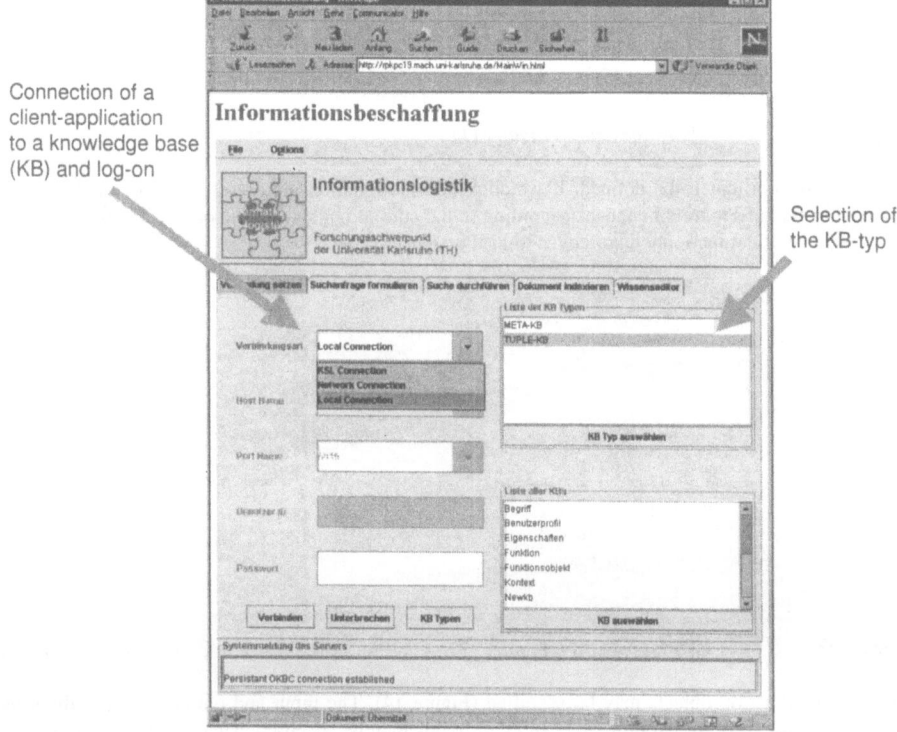

Figure 11:   *Establish a Client Connection*

In order to word a search inquiry, the user must change to the search register (Figure 12). Here, a user profile could be specified. User profiles are a possibility to narrow the search area and, when given, it accelerates the query process in the knowledge base. Generally said, the more information are provided in the inquiry, the more information may be found directly.

The task at hand is from Engineering => Mechanical Engineering => Design => Standard Part

Define profile, e.g.
Engineering
=> Mechanical Engineering
=> Design
=> Standard Part

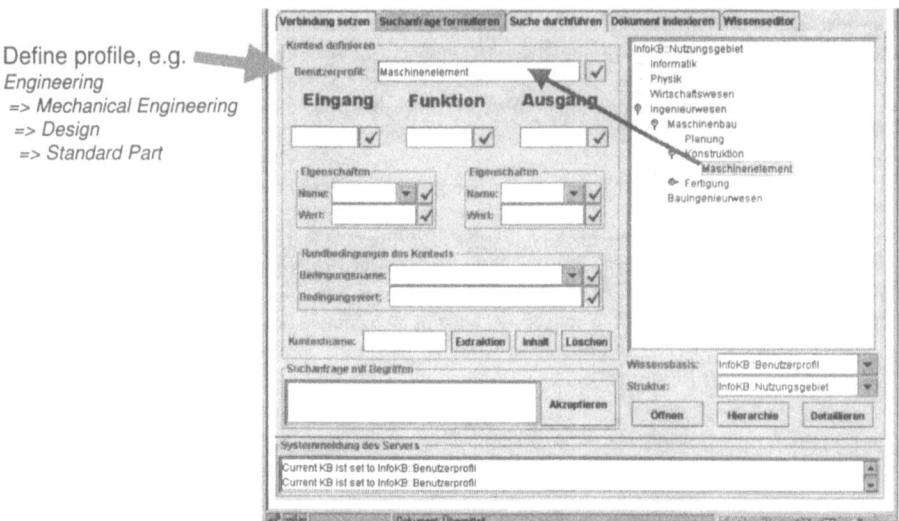

Figure 12:   *Word the Search Inquiry (1)*

Now the function has to be defined. If we click into the function input area then the window at the right hand side of Figure 13 changes according to the current context. In the example, a torque should be passed, i.e., the functional hierarchy is to lead by means of pass.

Define function,
e.g. *PASS*

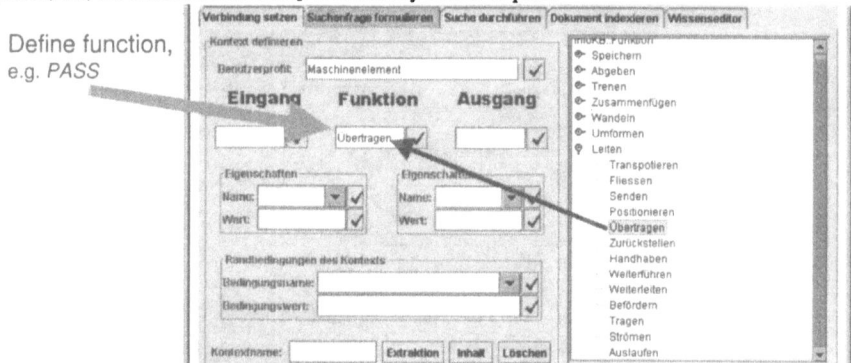

Figure 13:   *Word the Search Inquiry (2)*

Finally, the function objects may be specified (Figure 14). The input and output object is the same, namely a torque. Additional properties and constraints may be specified. They can be qualitatively or quantitatively specified. Here a small torque is to be passed.

Also visible in Figure 14 is the result of the inquiry. The answer to the question: How can a small torque be passed? Is: A key connection passes a small torque.

Define function objects (input/output), e.g. *Torque*

Additional properties and constraints for input/output, e.g. qualitative => *small* (fuzzy!!!) quantitative => *T <= 50 Nm*

Result of the inquiry, *here: key as connection*

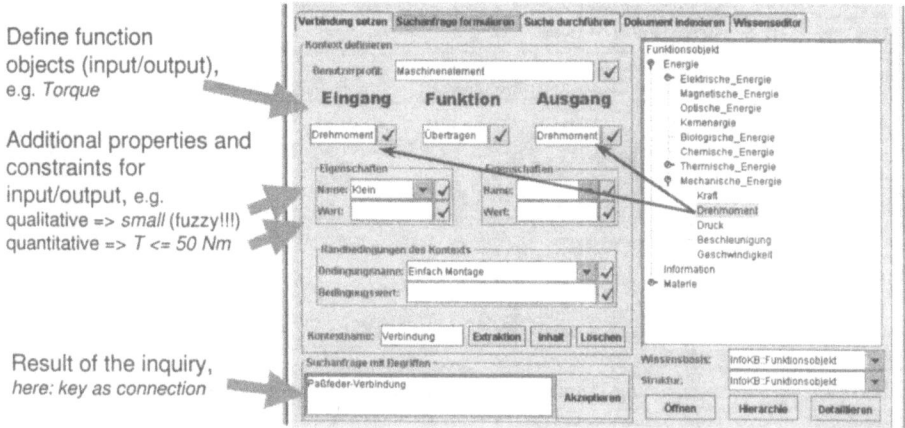

Figure 14:   *Word the Search Inquiry and Show the Result*

Beside the domain of mechanical engineering, functional aspects are significant in all technical and scientific domains. The development of new medicaments in the domain of chemistry/pharmacology may be a good example. The development of new medicaments is based on known cause/effect structures which are selected according to functional viewpoints, e.g. lower blood pressure, or improve to fall asleep [22].

3.3   Further Considerations – Formation of a Semantic Framework

The system for the automatic classification of products as well as the function based information retrieval system are filing systems which need the analysis of the available data in order to synthesise the required information. This process is generally called classification. Roughly spoken the strategy of the classification process is always very similar. It starts with the determination of the products, components, or items to be classified, i.e. what should be classified. Next, all possible distinguishing marks of the considered items are to be gathered. They are investigated according to their clarity, unambiguousness, and period of validity. Appropriate to the considered context the most important characteristics have to be chosen and determined. Then the instances of the properties of the items to be classified have to be determined, i.e. how are the instances mapped onto the properties. After these steps, the actual filing system can be build.

However, for each different context the classification schemes have to be created or modified and each filing system needs its own maintenance. Another aspect is the ability of understanding how the schemes were build since the context is usually kept implicit and hence not comprehensibly for people from different departments or domains. Also very often, people use the same terms and notions but the concepts being meant are totally different due to the misunderstanding of the considered context and hence the misinterpretation of the semantics. Figure 15 tries to illustrate the encoding, transfer, and decoding process for interpersonal communication.

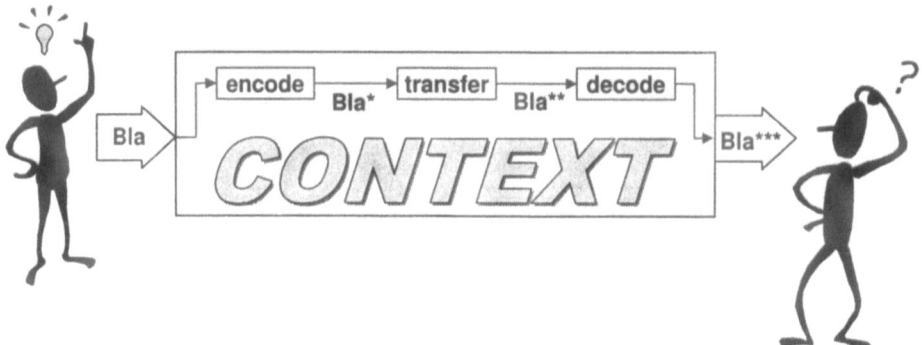

Figure 15:   *Encoding, Transfer, and Decoding Process for Interpersonal Communication*
                         [30]

Those disadvantages result in the demand for a *semantic framework*. A semantic framework is a system which is used to create and gather information models and knowledge units which represent a distinct view onto data and the applicability of the data represented in different representation systems. It is a system for the formal representation and (re-) use of notions and concepts with an explicit context, i.e. it supports the mapping of items of the universe of discourse with its distinguishing features onto notions and concepts with its features.

More precisely, it leads to a structured definition of terms and their distinctive features (e.g., a single part is a part that is not further decomposable and that is used as such) and to group terms into classes. It allows a structured definition of relations between objects, i.e., terms and their properties or attributes, respectively. Relations are predefined semantically rich relations, such as hierarchies, which are specialised according the considered semantics. Based on objects and their holding relations, a structured definition and distinction of theories will be provided. Such theories are specialised terms, objects, and relations in order to define a distinct context. For instance, when is a tool considered as a product and when as a tool?

Benefit of such a semantic framework is the reduction of ambiguous terminology through a sharp definition of terms, objects, relations, and theories regarding contexts (remember Figure 15). Based hereon, the system may be used for the re–use of theories for (complex) classification systems, conceptual data base development, and conceptual software development, etc.

Another aspect is to merge different (arbitrary) data sources through the complex and flexible formation of object classes and theories. For instance methods for data investigation may create varying views onto the data and allow the discovery of similarities based on theories. The function based information retrieval system, for instance, investigates informal data, such as text files, in order to find exact or partial matches.

As a vision, it might be serving as a general interface to access data, information, or to launch the appropriate representation or presentation systems.

In the following, the basic structure of the semantic framework will be briefly explained. The system is still under development.

The formation of the semantic representation system is based on a general knowledge model (KM) as described in CHAUDHRI et. all [5] (Figure 16). This KM allows the representation of constants, frames, slots (own and template), facets, classes, individuals (i.e. no set), and knowledge bases. According to CHAUDHRI et. all, it is a collection of constructs used in a wide range of representation systems.

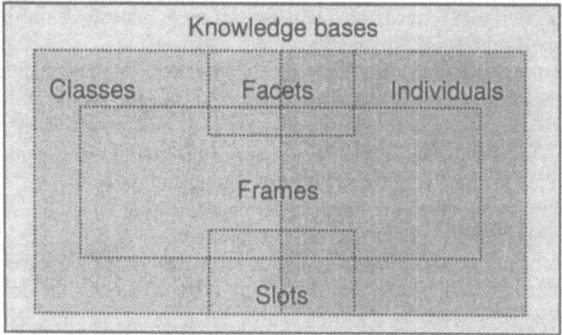

Figure 16:   *Open Knowledge Base Connectivity (OKBC) knowledge model [5]*

The semantic framework provides different representation strategies to handle different views and aspects on the data and to support different levels of granularity when building up a new model. These are associational representations (e.g., semantic networks and conceptual graphs), representations for structured objects (e.g., frames, schemes, units, and objects), formal logic-based (predicate logic, fuzzy logic), and procedural (i.e. production rules).

Therefore the representation formalism of Ontolingua/OKBC has been chosen. Ontolingua/OKBC serves as a manageable layer according to object oriented aspects as well as for predicate logic. Ontolingua as a formalism was developed in order to have an object oriented extension to the Knowledge Interchange Format (KIF) [15], [23]. KIF is a very expressive formalism which allows to define the semantics of a language. This may be used for the definition of other languages (it is not only a syntactical mapping but also a semantically one). OKBC serves as an application programming interface (API) for different client applications, such as the function based information retrieval.

The implementation of Ontolingua/OKBC and KIF was done with Common Lisp (see [5]) which is very flexible and allows to define further language layers. Advantage of this formalism is the neutral, application-independent representation. The representation layer is shown in Figure 17.

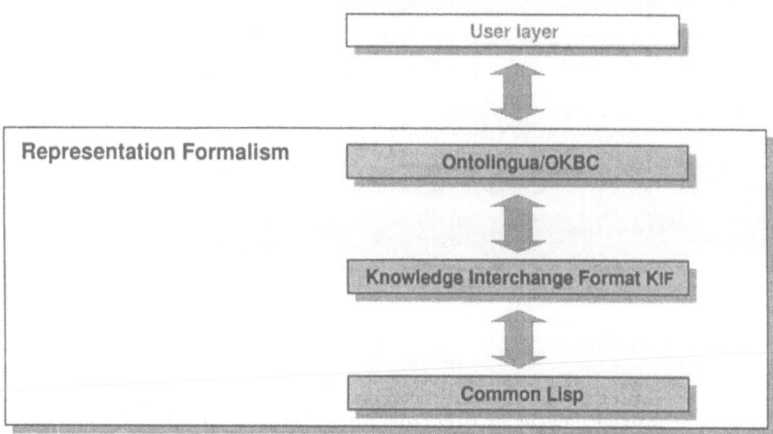

OKBC - Open Knowledge Base Connectivity by KSL-Stanford
(successor of Ontolingua Generic Frame Protocol GFP)

Figure 17:   *Formalism for Representation*

The semantic framework provides ordering and organising mechanism for the creation of new terms, concepts, and theories in terms of semantically rich relations. In general there are two different types of relations, i.e. hierarchical and non-hierarchical relations.

Hierarchical relations are roughly subdivided into abstracting or generalisation/specialisation relations (i.e. supertype/subtype) and relations for aggregations (i.e. part-of).

Non-hierarchical relations are subdivided into sequential relations and pragmatic relations. Sequential relations are predecessor/successor relations (e.g. process chain for manufacturing), causal relations describing causal/effect contexts, transmission relation (e.g. emitter/receiver), genetic relation (e.g. producer/product), manufacturing relation (e.g. material/product, instrumental relation (tool/application of the tool), and functionally relation (argument/function).

Pragmatic relations are those which are thematically connected but which are nor hierarchical either sequential relations. An example may be the relation car/street.

For the creation of new terms, syntactical relations are taken into account. Syntactical relations such as determination, conjunction, disjunction, and integration of terms are for the definition of new terms and the recognition of terms given by user interaction. For instance, possible syntactical relations for man and woman are for the determination virago/transvestite, for conjunction it is a hermaphrodite, for the disjunction it is an adult, and for the integration it is a (married) couple.

This way of organising terms, concepts, and theories enables the indexing of the knowledge base of the system for inferring tasks done by the system. It also facilitates the definition of new, composite and, complex terms and concepts since considered facets (if represented in the system) may be combined in order to express the intention and semantics of new terms. In other words, Basic terms and relations provide an alphabet, a syntax, and semantics, i.e., what may be expressed and how can it be expressed, and what is the basic meaning of the expressed statements. The pragmatically aspect of the semantics, i.e. the misinterpretation by humans which leads to misunderstanding should be minimised or eliminated, respectively. Specialising basic terms and relations towards distinct domains enables the formal definition of a branch specific view onto items of the world. Here, general concepts and theories are represented. Company specific terms, relations, and assumptions are used to represent knowledge units oriented at problems and tasks at hand. The representation layer of the semantic frame system is shown in Figure 18.

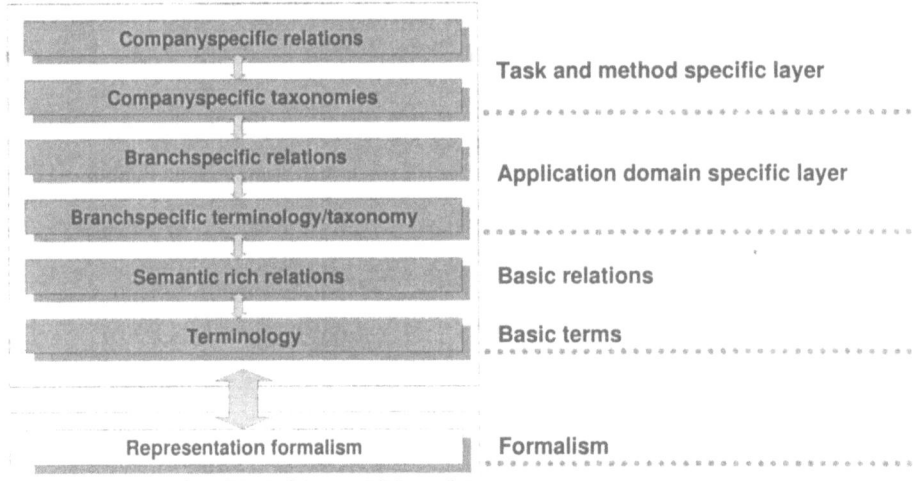

Figure 18:   *Realisation of the User Layer for the Semantic Framework*

The semantic framework is currently developed at RPK. The ideas were coming from the experience gained through the modelling of complex structures, such as the integrated product and production model, or the application protocol for automotive design within STEP (ISO-10303 Part-214). Here, a lot of problems arose due to the fact of different interpretation of terms and notions and on concepts and theories. The semantic framework is an approach to reduce the complexity of modelling tasks and to increase the re-usability of components of a model for different client applications.

4   Conclusion

In the last section, a first conclusion was already given, so a short résumé will be drawn.

Living data are data that are in use or that are (re-) useable for human beings and machines. In other words, living data are information with a distinct semantics in a considered context, that are valid within a distinct period, that are available and accessible, and hence, that are (re-) useable for human beings and machines.

It has been shown that classification and classifying is a central issue for handling and re-use of complex data. Therefore, terminology has to be sharp defined, i.e. non-ambiguous, and the semantics and context must be clear in order to enable usability by different human beings. On the other hand side, correct understanding of the data and their semantically relations may speed up modelling and standardisation activities: If humans do not understand terms, terminology, and theories about, how should they tell it a computer?

Data must be useable by machines, i.e., the creation and re-use of theories for analysing and ordering of complex data and the merging of different data sources in order to create different views must be possible. Therefore the account to the internal knowledge model of the application must be given. This results in the demand for open software architectures.

5   Acknowledgement

The system for the *Automatic Classification of Products* is funded by Federal Ministry of Education, Science, Research, and Technology (BMBF) and it is developed by Alexander Staudinger, Joerg Weisskopf, and Matthias Wunsch (RPK). Industrial requirements are coming from AUCOTEAM, Berlin (production automation, measurement- steering- and regulation- technology), Eigner & Partner, Karlsruhe (CADIM PDM), GHH Borsig Turbomaschinen, Berlin-Oberhausen (compressor and turbines), and Kloeckner-Moeller, Bonn (electronic components for switching and distribution). All locations are in Germany.

The system for the *Function based Information Retrieval* is funded by the Ministry for Science, Research, and Art of Baden-Wuerttemberg, Germany, and it is developed by Rainer Ostermayer, Christian Klimesch, and Jing Fu (RPK) within the project *Information-Logistics for Branch-spanned Co-operation*.

The basic research work of the *Semantic Framework* was funded by German Research Community (Deutsche Forschungsgemeinschaft – DFG) in the project *Computer Aided Representation of Engineering Knowledge for an Efficient Re–Use of Technical Solutions* [21].

Acknowledgements also to Knowledge Systems Laboratory, Stanford University, USA and SRI International, Menlo Park, USA for the development of Ontolingua and especially for OKBC.

6   References

[1] Abeln, O.: Vom CAD-Arbeitsplatz zur Konstruktionsleittechnik (Teil 1); CAD-CAM Report Nr. 5; Dressler Verlag, Heidelberg; 1991.

[2] American National Standards Institute (ANSI): ANSI Y14.26M-1981: Digital Representation for Communication of Product Definition Data IGES (1.0); Herausgeber: American National Standards Institute (ANSI); 1981.

[3] Bock, H.-H.: Methoden und Probleme der numerischen Klassifikation; In: Dahlberg, I.; Dahlberg, W; (Hrsg.): Prinzipien der Klassifikation; Proceedings der 1. Fachtagung der Gesellschaft für Klassifikation e.V., Bd.1 (SK1), 1977.

[4] Bowman, Judith S.; Emerson, Sandra L.; Darnovsky, Marcy: The practical SQL Handbook - Using Structured Query Language; 3. Auflage; Addison-Wesley Developers Press, Reading (Massachusetts), etc.; ISBN 0-201-44787-8; 1996.

[5] Chaudhri, V. K.; Farquhar, A.; Fikes, R.; Karp, P. D.; Rice, J. P.: Open Knowledge Base Connectivity 2.0.3; Knowledge Systems Laboratory, Stanford University (USA); Report Nr. KSL-98-06; 09.04.98.

[6] Cooper, W.S., Gey, F.C.: Experiments in the Probabilistic Retrieval of Full Text Documents; In: The Proceedings of the Third Text Retrieval Conference (TREC-3), 1994.

[7] CSC Ploenzke AG: Life Cycle Management – Improving Products and Processes for Tomorrow's Markets (7. Internationaler Kongreß); Konferenz: Kongreßzentrum Mainz, 05.-07.05.; Organisation: CSC Ploenzke AG; Herausgeber: CSC Ploenzke AG; Wiesbaden (Kreuzberger Ring 62, D65205 Wiesbaden); 1998.

[8] Dahlberg, I.; Dahlberg, W.: Prinzipien der Klassifikation; Proceedings 1. Jahrestagung der Gesellschaft für Klassifikation e.V., Studien zur Klassifikation, Bd. 1 (SK 1), Münster, 4. Juni, 1977.

[9] Degens, P.O.; Hermes, H.-J.; Opitz, O.: Die Klassifikation und ihr Umfeld; Proceedings 10. Jahrestagung der Gesellschaft für Klassifikation e.V., Studien zur Klassifikation, Bd. 17 (SK 17), Münster, 18.-21. Juni, 1986.

[10] Deutsche Presseagentur (DPA) München: Milliardenverluste durch chaotische Datennetze; Der Tagesspiegel (Wirtschaft); ab S. 14; Berlin; 04.08.98.

[11] DIN - Deutsches Institut für Normung e.V.: DIN 66301: Rechnerunterstütztes Konstruieren - VDA-FS; Herausgeber: DIN - Deutsches Institut für Normung e.V.; Beuth Verlag GmbH, Berlin; 1984.

[12] Dirzus, E.: Reorganisation des Informationswesens bei integrierter Auftragsabwicklung in Maschinenbauunternehmen - Voraussetzungen, Planung, Durchführung. Dissertation, RWTH-Aachen, 1972.

[13] Ehrlenspiel, K.: Frühzeitige Kostenbeeinflussung durch Produktkosten-Controlling und Simultaneous Engineering; Nr. 6; S. 313-320; 1995.

[14] Friedemann, Th.: EDM: Stand der Technik und Trends; CSC Ploenzke AG, Competence Center Industrie; 1995.

[15] Genesereth, M. R.; Fikes, R. E.: Knowledge Interchange Format Version 3.0 Reference Manual; Interlingua Working Group of DARPA Knowledge Sharing Effort; Bericht Nr. Logic-92-1; Computer Science Department, Stanford University, Stanford, California (USA); 6/1992.

[16] Grabowski, H. (Hrsg.): Rechnerintegrierte Konstruktion und Fertigung von Bauteilen; Sonderforschungsbereich 346, Universität Karlsruhe; Kolloquium am 30. Juni 1998; 1998.

[17] Grabowski, H. (RPK); Dillmann, R. (IPR); Kohler, N. (ifib); Scherer, R. J. (IMBB); Schmid, D. (IRF); Spath, D. (wbk); und Mitarbeiter: Informationslogistik für unternehmens- und branchenübergreifende Kooperation – Antrag auf Weiterführung der Arbeiten des "Schwerpunktes Informationslogistik"; Schwerpunkt Informationslogistig an der Universität Karlsruhe; Antrag Nr. 970421; Karlsruhe; 21.04.97.

[18] Grabowski, H.; Lossack, R.-S.; Klaar, O.; Weißkopf, J.: Klassifikation als Werkzeug zur effizienten Lösungssuche auf Basis von Anforderungen; in: Fertigungsgerechtes Konstruieren – Beiträge zum 10. Symposium; Konferenz: Lehrstuhl für Konstruktionstechnik, Friedrich-Alexander-Universität Erlangen-Nürnberg, Schnaitach, 15.-16.10.; Herausgeber: Prof. Dr.-Ing. Meerkamm, Harald; 1999.

[19] Grabowski, H.; Weißkopf, J.: Metadatenkonzept zur strukturierten Erfassung von Produktmerkmalen für die automatische Produktklassifikation; in: Fertigungsgerechtes Konstruieren – Beiträge zum 9. Symposium; Konferenz: Lehrstuhl für Konstruktionstechnik, Friedrich-Alexander-Universität Erlangen-Nürnberg, Schnaitach, 15.-16.10.; Herausgeber: Prof. Dr.-Ing. Meerkamm, Harald; 1998.

[20] Grabowski, H; Rude, S; Liu, C.; Hain, K.: Finden von Wiederhollösungen auf Basis von Informationsrekonstruktion und automatischer Klassifikation; 9. Forschungsseminar der Hochschulgruppe Arbeits- und Betriebsorganisation (HAB), St. Gallen, 1996.

[21] Grabowski, Hans; Ostermayer, Rainer: Rechnerunterstützte Darstellung von Ingenieurwissen für die effiziente Wiederverwendung von technischen Lösungen; Institut für Rechneranwendung in Planung und Konstruktion (RPK), Uni-Karlsruhe; DFG-Ergebnisbericht; Karlsruhe; 1998.

[22] Grabowski, Hans; Ostermayer, Rainer; Klimesch, Christian: Function Based Information Retrieval; in: Integrated Product Development - IPD 98 - 2nd International Workshop; Konferenz:

Lehrstuhl für Maschinenbauinformatik, Uni Magdeburg, Magdeburg, 17.-18.09.; Herausgeber: Prof. Dr.-Ing. Sándor Vajna; 1998.

[23]   Gruber, Thomas R.: Ontolingua – A Mechanisim to Support Portable Ontologies; Knowledge Systems Laboratory, Stanford University (USA); Bericht; Palo Alto California, USA; 6/1992.

[24]   Hain, Karl: Automatische Gewinnung von Merkmalen und Klassifizierungseigenschaften für Produkte auf Basis eines integrierten Produktmodells; Reihe: Forschungsberichte aus dem Institut für Rechneranwendung in Planung und Konstruktion (RPK) der Universität Karlsruhe; Herausgeber: Prof. Dr.-Ing. Dr. h.c. H. Grabowski; Band 97 Nr. 4;1. Auflage; Shaker Verlag, Aachen; ISSN 0945-5787; ISBN 3-8265-3236-8; Dissertation RPK Uni-Karlsruhe; 1997.

[25]   Inscore, Jim: Introducing Java IDL; Sun Microsystems, Inc.; Tutorial; 1998.

[26]   Knorz, G.: Automatische Indexierung; In: Hennings; Knorz; Rinicke; Schwandt: Wissensrepräsentation und Information Retrieval, Universität Potsdam, 1995.

[27]   Knorz, G.: Information Retrieval Anwendungen; In: Zilahi-Szabo (Hrsg.): Kleines Lexikon der Informatik und Wirtschaftsinformatik, Oldenbourg-Verlag, München, Wien, 1995.

[28]   Knorz, G.: Neue Aspekte der Informationsverarbeitung – Inhaltserschließung und Retrieval; In: Bullinger, H.-J. (Hrsg.): Dokumentenmanagement bei öffentlichen Dienstleistern, IAO-Forum, 1995.

[29]   Knorz, G.: Technologie und Methodik von Informationsspeicherung und Retrieval; In: AWV, FMI (Hrsg.): Bits & Bytes, Mikrofilm und Papier - Informationsspeicher der Gegenwart und Zukunft; Orgatec-Forum 1994, Köln.

[30]   Lenat, Doug: The Dimensions of Context-Space; Cycorp; Report; 3721 Executive Center Drive, Suite 100, Austin; 08.10.98.

[31]   Lu, Stephen C.-Y.: Die Generation nach regelbasierten Expertensystemen - Die Generation nach regelbasierten Expertensystemen - Stand der Forschung in technischen Anwendungen der künstlichen Intelligenz in den USA; in: Erfolgreiche Anwendung wissensbasierter Systeme in Entwicklung und Konstruktion; Konferenz: VDI, Heidelberg, 7.-8.10.; Reihe: VDI Berichte; Nr. 903; VDI-Verlag, Düsseldorf; 1991.

[32]   Maderholz, R.: Umsetzung von Benutzeranforderungen an die CAD-Normteiledatei - Funktionsgruppen, Anwenderoberfläche, Auswahllogik; In: Senk, G. (Hrsg.): Referatensammlung NormCAD 92, DIN e.V., Beuth, Berlin/Köln, 1992.

[33]   N.N.: Overview about Cyc; Cycorp Inc.; 3721 Executive Center Drive, Suite 100, Austin, TX 78731; 3/1999.

[34]   N.N.: White Paper about CoBrain; Invention Machine Corporation (IMC); Invention Machine Corp. World Headquater, 133 Portland Street, Boston, MA 02114-1722, USA; 1999.

[35]   Opitz, H.: Die richtige Sachnummer im Fertigungsbetrieb; Eine Basis für Rationalisierungsmaßnahmen im Rahmen der Auftragsabwicklung; Essen: Girardet, 1971.

[36]   Opitz, H.: Werkstückbeschreibendes Klassifizierungssystem (Teil 1); Ver-schlüsselungsrichtlinien und Definitionen zum werkstückbeschreibenden Klassifizierungssystem (Teil 2); Essen: Girardet, 1966.

[37]   Owen, Jon: STEP – An Introduction; Reihe: Product Data Engineering; Herausgeber: Owen, Jon; Information Geometers Ltd., Winchester (UK); ISBN 1-874728-04-6; 1993.

[38]   Schütze, H.; Pedersen, J.O.; Hearst, M.A.: Xerox TREC-3 Report: Combining Exact and Fuzzy Predicators; In: The Proceedings of the Third Text Retrieval Conference (TREC-3), 1994.

[39]   Späth, L.: Kreatives Management der Unternehmensführung; In: Zahn, E. (Hrsg.): Mit Kreativität die Zukunft meistern; Stuttgart, Schäffer-Poeschel; 1995.

[40]   VDI-Gesellschaft Konstruktion und Entwicklung: Blatt1: Erstellung und Anwendung von Konstruktionskatalogen, VDI-Verlag Düsseldorf, 1982; Blatt 2: Konzipieren technischer Produkte, VDI-Verlag Düsseldorf, 1977.

[41]   Warnecke, H.-J.; Becker, B.-D.: Strategien für die Produktion im 21. Jahrhundert; Fraunhofer-Gesellschaft zur Förderung der angewandten Forschung; Bd. 1: Hauptbericht, 1994.

[42]   Wiendahl, H.-P.: Funktionsbetrachtungen technischer Gebilde - Ein Hilfsmittel zur Auftragsabwicklung in der Maschinenbauindustrie; Dissertation, TH Aachen, 1971.

Enabler Technologies
for Future Engineering Processes

Siegmar Haasis, Alfred Katzenbach

DaimlerChrysler AG, Research Center, Ulm, Germany
Siegmar.h.Haasis@daimlerchrysler.com

Abstract: After defining the requirements of future engineering processes enabler technologies are explained. They are divided into the basic technologies global product data management, distributed system platforms, generic product data model and integration architecture and into several application oriented technologies like feature technology or agent technology. These technologies are the base to cope with the requirements of the future.

1 Introduction

1.1 General Improvement of the Automotive Industry

At the end of the 20th century globalization has lead to a worldwide competition in nearly all branches. Especially in the automotive industry there is a keen competition. In mature markets an increasing over-capacity can be found and new markets can only be developed by lower prices and increasing local content. Time to market has decreased significantly during the last decade. So, for innovative companies it is very important to develop new innovative products and to bring them on the market at the right time. Time and by this way engineering processes are becoming more and more important for the competitiveness on the world market. This leads to a concentration process of original car producers and system partners and an enforced need to proceed in globalization of development and manufacturing of vehicles. Furthermore the keen competition requires a differentiation to competing car producers by individual and innovative products and by attractive prices.

1.2 Requirements for Product Engineering

As a result of this situation the following requirements for product engineering arise:
The creation of attractive products must follow the current lifestyle. The time to market from product definition to product launching has to be reduced.
Product innovation is the critical success factor especially for the premium market. Thus research results have to be introduced into the products rapidly.
Attractive prices are one key for the success in the volume market. Economy of scales can be realized by the modularization of product components.
Extra cheap products create advantages in third world countries. So, new creative ideas for inexpensive products and manufacturing processes have to be developed and product concepts have to allow an increasing local content.

1.3 Collaboration with Suppliers

As there is a keen competition on the world market car producers are focusing on their core business. By this way the cooperation of the car producers with their suppliers becomes very important and changes from part suppliers towards system suppliers are going on. This is one reason, why the globalization of the supplier market is growing even faster than the OEM (Original Equipment Manufacturer) market. To realize economy of scales the same components of one system supplier are integrated in products of different car producers. It has to be ensured that the suppliers are embedded in the development process like the OEM internal partner.

2 Current Requirements

2.1 Vision for Future Engineers

In the future engineers should be able to consider all different topics which are important within the life cycle of a product i.e. requirement analysis, development (functional aspects, strength and durability, ...), planning (costs), manufacturing, configuration and assembly, marketing and sales, use, maintenance and recycling. This finally enables the development of products "straight to target".

2.2 Requirements for the IT-Systems

The support of engineering processes by IT-systems is becoming more and more a key for success on a competing market. In order to achieve this IT-systems have to fulfill the following requirements:
single intuitive and context sensitive user interface,
high performance and stability,
system independent and consistent availability of data,
associativity between contents of data,
high security of data,
flexible communication with engineering partners just in time and worldwide.

2.3 Tendencies of the IT-Supplier

Among the IT-suppliers of CAx-systems, of EDM-systems and of ERP-systems a concentrating process can be seen. However, their offered main systems support only main stream processes. A huge customizing effort is necessary to change standard systems to proprietary systems. A lack of nearly all systems is the support of concurrent and simultaneous collaboration worldwide and between companies. The customer requirements force IT vendors to open their system architectures, but openness of their systems is against their own competitive interests.

2.4 IT Support for the Development Process

To ensure that IT supports the development process correctly the following topics have to be considered:
Current design methods are focusing on the design and are not considering the continuos data flow following the design.
The current CAx-systems are supporting the change from part oriented to assembly oriented development.
The opportunities of modern information technologies are only used by a small percentage.

2.5 Changes in the Engineering Process

Changes in the engineering process lead to the following requirements:
The generation of product describing data has to be done comprehensively and to be consistent within a network of connected systems.
The experience of physical tests and practical use of the vehicles have to be represented in the entire data set, which has to be available across the whole process chain.
There must be the possibility to visualize and simulate every single configurable product.
The comprehensive engineering process has to be understood across company borders and together with system suppliers.

3 Enabler Technologies

Even today some technologies are available which help us to cope with the requirements of the future. These technologies can be divided into basic technologies and application-oriented technologies (figure 1). Some of them are described in detail [1,2].

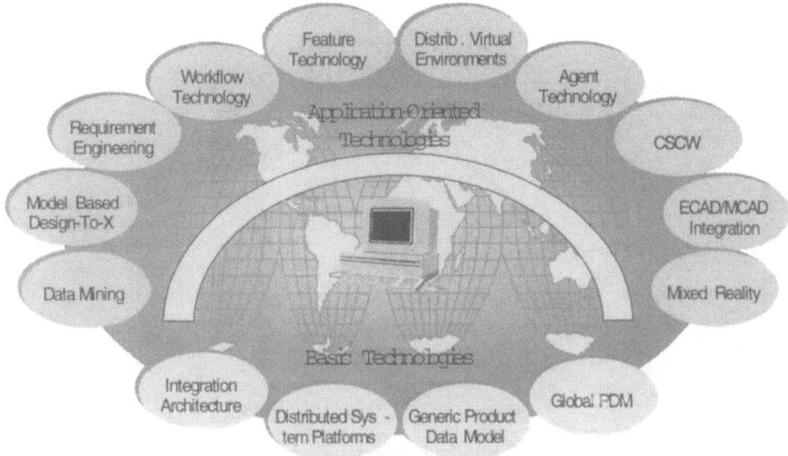

Figure 1: Examples of enabler technologies

3.1 Basic Technologies

Distributed System Platforms
Communication Systems and Distributed Systems are the foundation of distributed engineering processes and distributed Virtual Reality. In addition, multimedia communication allows to communicate with audio and video over the same telecommunication system.
High-speed communication systems with Quality of Service (QoS) guarantees and distributed system platforms are the enabler technologies for distributed engineering processes and distributed Virtual Reality. Furthermore they are the base for a seamless integration of synchronous voice and video communication in the engineering applications.
Distributed System Platforms based on CORBA are used for the control and management of the applications. High-speed TCP/IP-based communication with QoS guarantees are used for the transfer of bulk data. Multimedia applications such as video conferencing and CSCW tools are seamlessly integrated into the application environment.

Generic Product Data Model
Each existing PDM system has its own data model which is - in most cases - very different from the data models of the other systems. Even in overlapping areas (e.g. part identification) the data models of the PDM systems differ. An enterprise wide data model does not exist. Thus the communication between systems requires a common language, i.e. a common data model.

A common product data model which can be extended in a flexible manner according to the requirements is the prerequisite of integration and interoperation. The role of each PDM system becomes much clearer if its data model is mapped to the common data model.
The STEP data model AP214 (the adapted data model "MB-PDM") is the core product data model. Product data exchange processors are and shall be based on that data model. An integration architecture supporting data sharing shall also be based on that data model.

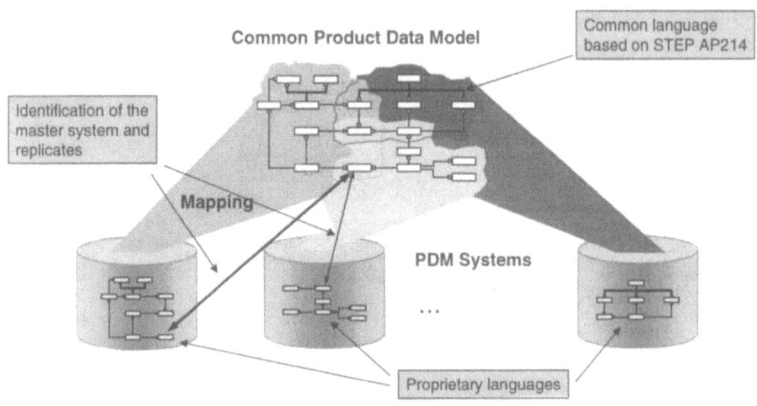

Figure 2: Generic Product Data Model

Global Product Data Management
Today engineering resources are enterprise wide distributed. To use them efficiently we need technologies for a common and simultaneous development of products within world wide distributed engineering teams. Therefore global distributed product life cycle information (product data, tool info, processes, ...) from the pre sales up to the after sales have to be managed in a distributed and heterogeneous environment.
Technologies for management of a distributed environment have to be implemented to achieve system stability, data consistency, high performance data access, backup- recovery-mechanisms etc. which will impact the engineering process. Data and processes based on a common product life cycle model and a common integration architecture (partners, suppliers) have to be fully integrated, which leads to harmonization and standardization of technology, tools and methods.
One example is the implementation of common-core at DaimlerChrysler Rail Systems. It contains the definition of an enterprise wide product life cycle information model including product data, user rights, tools, processes, methods, procedures, knowledge etc. Another example is the ongoing realization in the project ROCHADE at DaimlerChrysler Aerospace where processes are described to enable global engineering (process steering and control; management of user rights, authority; user defined access mechanisms). One last example is the implementation of individual, user defined views on the common data which where realized in the electrical process chain at DaimlerChrysler Aerospace – Airbus.

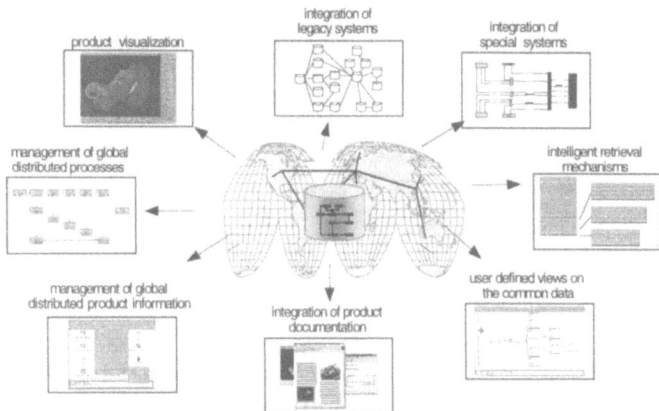

Figure 3:   *Global Product Data Management – during product life cycle, world wide access to all product*
life cycle information

Integration Architecture

Today, the engineer is supported in a system-specific but not task-oriented way. Task-oriented IT-support requires a single system image so that heterogeneity, data redundancy and distribution of functionality is transparent to the engineer.

The impact on the engineering process is a better support of task-oriented processing, a reduction of information retrieval time (faster response time to access the relevant information), less PDM systems to operate on (incl. know-how) to get the relevant information, an improvement of the quality of work because of the higher availability of information and a better support in introducing new PDM systems.

The prototype "EDM-InfoManager" is ready for pilot testing. Potential application areas are the integration of PDM-systems from DaimlerChrysler Auburn Hills and DaimlerChrysler Stuttgart, access from systems of the sales division to the engineering systems and (external) access from suppliers to the PDM systems.

3.2   Examples of Application-Oriented Technologies

In the following chapter some examples of interesting application-oriented technologies are explained.

Feature Technology

A feature is a partial form or a product characteristic that is considered as a unit and that has a semantic meaning in design, process planning, manufacture, cost estimation or other engineering disciplines. Such a meaning is useful for reasoning about the product. Feature technology is necessary to accelerate product modeling and to improve process chains [3].

Design engineers are thinking in terms of functional objects like i.e. bearings. However current CAD-systems are focused on geometrical aspects only. Thus the development of a feature editor and feature-based applications had to be done to describe a product in an object-oriented way. This enables a description of the product based on design features according to engineers' terms of thinking. Features provided by a library are supporting a standardized and manufacturable product design.

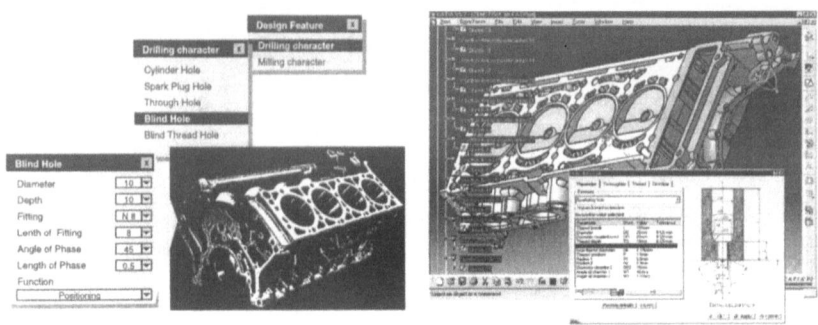

Figure 4: *Feature-based design*   **Figure 5:** *Feature-editor based on CATIA*

The information captured in features can be transferred to process planning and manufacturing. The content of a feature may change depending on the point of view but a lot information can be shared in design, process planning and manufacturing view. Feature technology helps us to capture more information of process and manufacturing knowledge. Operations and tools can be assigned to a feature and resources (tools and machines) can be allocated. This leads to a higher transparency during the modeling of process and resource and finally to a complete digital description of product and process. So, feature technology is the base for the implementation of a bi-directional backbone to connect product description, process planning and resource optimization [4].

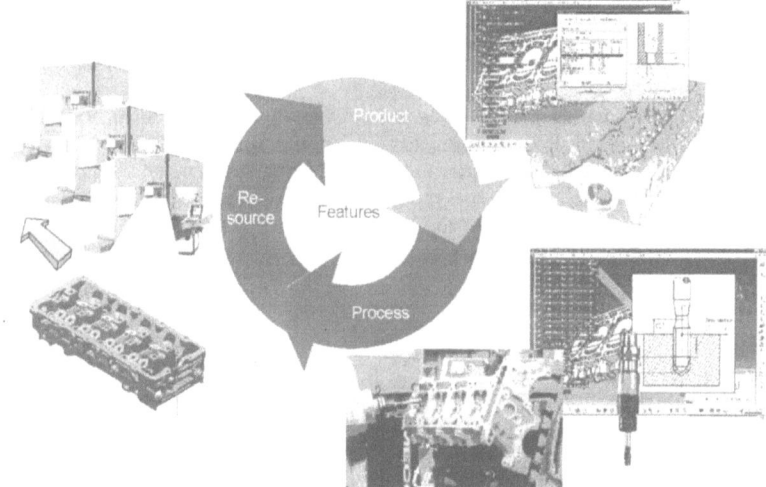

Figure 6: *Feature technology – Connection of product, process, and resource*

A prototype of a feature editor within the CAD-system CATIA V5 for the feature-based design and a prototype called Object-NC (CATIA V5 application) for feature-based manufacturing are provided in cooperation with the DaimlerChrysler plant in Untertürkheim. It is planed to employ both in the near future. Further feature-based applications for design, process planning and resource optimization (cost calculation system, design checker, technology assistant, ...) are in development.

Agent Technology

Agent technology is a new software technology. Agents are autonomous objects which are responsible for a specific task by themselves. They coordinate with other agents regarding their goals. An agent for example can be a process object (work piece, transport unit, manufacturing station) which takes care for its next process step. The work piece agent can then look for a free machining center which is able to fulfill the next machining operation.

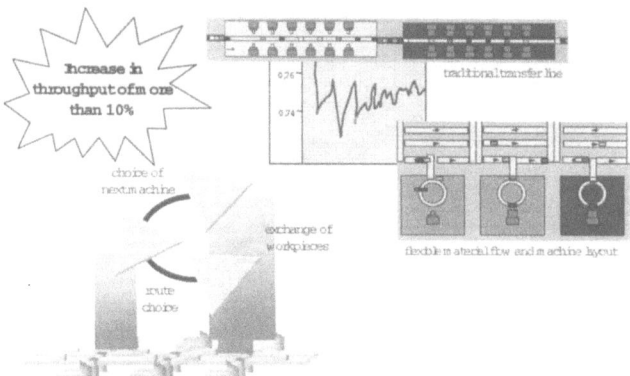

Figure 7:   *Agent technology – Controlling changes and disturbances in manufacturing*

The advantage of agent technology is the possibility of adapting the manufacturing process to changes and disturbances quickly. Changes are due to e.g. modifications in design details of work pieces, varying slot sizes of variants etc. Disturbances are caused e.g. by machine or transportation system breakdown, supply chain interrupts or delays etc. By combining agent and feature technology late changes in design details can immediately be executed in the manufacturing process. The difficulty of demand prediction can be met by dynamic adaptation of the manufacturing unit to varying slot sizes of model series and model variants.

An agent-based control software for a prototype plant is realized in the DaimlerChrysler plant in Untertürkheim (cylinderhead production).

Distributed Virtual Environments

Engineering knowledge becomes more and more spread over world-wide distributed development departments. Thus tools for distributed concept design conferences are an important requirement. For the distribution and acquisition of information and knowledge intranets and internet can be used as a framework and backbone. Furthermore the acceptance of virtual reality as fast and generic 3D-evaluation environment increases.

A close connection of Distributed Virtual Environments to the engineering processes reduces the errors and costs for physical mockups. Distributed 3D-visualization conferences reduce the number of very expensive errors in the early development phase. In addition the distribution of information and visualized interaction even over low bandwidth networks save travel costs.

To integrate this technology into engineering processes the conversion gap between CAD-data and VR-data has to be closed by linking EDM/PDM systems with VR environments. A Request Broker architecture for the distribution of data and interaction commands within shared virtual reality environments has to be provided. Tools for interaction, evaluation and navigation and standards for distributed virtual engineering conferences have to be developed.

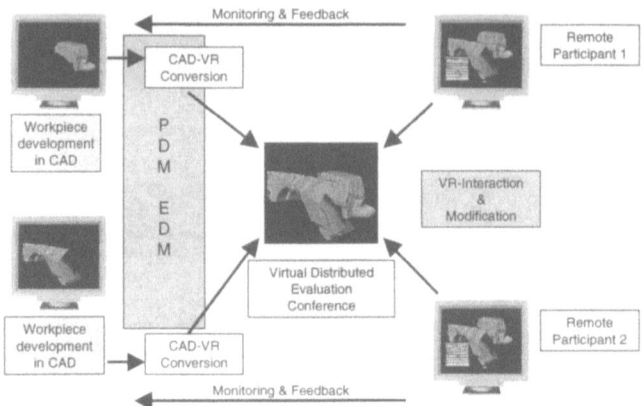

Figure 8:   *Distributed Virtual Environments*

Workflow Technology

Workflow technology is deployed to achieve transparent processes and to reduce time-to-market. Today's Workflow Management Systems do not offer adequate support for engineering processes and do not enable real concurrent engineering.

By the use of work flow technology engineering processes become more transparent, process data are monitored automatically, quality gates and milestones are controlled automatically and the coordination of process participants is improved which is one condition for concurrent engineering. Finally this leads to a reduced time-to-market with lower costs.

A new Workflow Management System called WEP has been developed and is ready for pilot testing. The system will be validated in a productive environment at DaimlerChrysler. Future implementations of WEP will use the EDM-Info Manager technology.

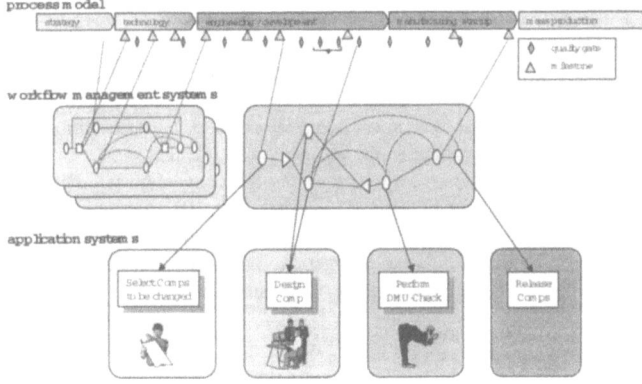

Figure 9: Workflow Technology

4   Conclusion

The explained technologies are not broadly used yet, but they show that there are ways to cope with the requirements of the future. Although the future cannot be predicted, it can be influenced and created by us. Issue of this presentation was the consideration of future prospects from different points of view, possible technological developments were emphasized.

To improve the company success, technologies but also processes and human behavior have to be developed and optimized entirely. This is the common challenge for vehicle producers together with IT system suppliers [5].

5 References

[1] Katzenbach, A.: International Engineering Processes of the Future. In Proceedings of Forschungsforum Universität Stuttgart "Wissenmanagement und schnelle Produktentwicklung", Fellbach, 12th November, 1999.

[2] Katzenbach, A.: Future Engineering Environment at DaimlerChrysler. In Proceedings of Daimler-Chrysler Corporate Technology Colloquium "Digital Product Development and Manufacturing", Wiesensteig, 5th to 7th May, 1999.

[3] Haasis, S.: Feature-Based Process Chain Within the Scope of Powertrain Engineering. In Proceedings of the 31st ISATA, Düsseldorf, 2nd to 5th June, 1998.

[4] Haasis, S.; Frank, D.; Rommel, B.; Weyrich, M.: Feature-basierte Integration von Pro-duktentwicklung, Prozeßgestaltung und Ressourcenplanung. In Proceedings of Information Technology in Design '99, München, 19th to 20th October, 1999.

[5] Balasubramanian, B.; Winterstein, R.: Challenges in the Digital World of Automotive R&D. In Proceedings of the 5th European Concurrent Engineering Conference, 26th to 29th April, Erlangen-Nuremberg, 1998.

CARMEN Technology: Towards a Component CAx System

Andrzej T. Janocha

CAxOPEN product development technology GmbH, Kaiserslautern
janocha@caxopen.de

Abstract: As broader parts of the product development process are becoming supported through computer aided technologies, the role of interoperability between heterogeneous CAx systems involved in this process is becoming more and more important. In the past, many efforts focused on defining formats for exchange of product data, e. g. IGES, STEP. This satisfied the sequential development process and the exchange of simple objects (like technical drawings) but is insufficient for parallel processes and exchange of complex objects. Instead of data formats, efficient mechanisms for tools integration and product data exchange are required. This paper describes embedding of CAx objects in other systems as one possible mechanism to achieve a better integration across different design stages.

1 Introduction

Interoperability, which is the ability to communicate the product data across different design and manufacture activities, becomes the key-issue because more and more parts of the product development process are becoming supported through computer aided technologies.

Until now, only the data view of the interoperability problem was considered. Many efforts focussed on defining and implementing formats for product data exchange. The IGES or the STEP/PDES activities are the best examples. With the STEP/PDES background, high- performance converters which translate native data representations into the standard representation and vice versa can be implemented. Nevertheless, important paradigms like parametric or history-based design are still not part of this standard.

On the other hand, concurrent engineering which integrates design, manufacturing and other processes to provide early manufacturing input into the design process, requires new access methods to product data. The product data flow changes from sequential and discrete into parallel and continuous. More and more design activities require exact design context on demand.

And last but not least, the system landscape becomes more and more heterogeneous. This is obvious, when we consider, that almost each activity incorporated into the virtual product development, for example CAE, uses dedicated tools from different specialised system suppliers.

As a consequence, a practicable solution of the interoperability problem should deal with all these dimensions and represent a *mechanism* rather than a *format* for product data exchange.

The following paper describes one possible approach to the interoperability problem, which considers both the product data modeling view as well as the data logistics view and the software system dimension. A case study of a multi CAD environment consisting of CATIA and Pro/ENGINEER CAD systems shows the implementation and potential of this approach.

2   Interoperability Issues

For a long time, interoperability between CAx systems has stood for product data exchange via neutral file formats like IGES or STEP/PDES. As the virtual product development integrates more and more process elements, other requirements on interoperability have emerged. The primary goal to exchange topologically and geometrically correct product data between systems still stands in the foreground. Additionally, a new appropriate product data flow and a better integration of the CAx software tools across the whole development process are required.

Product Data Exchange

The structure and content of CAD models has changed dramatically in the last years. It evolved from a set of simple geometric entities representing technical drawings (like lines, circles and several symbols) into complex topological structures including history and parametrics and representing solids or also other objects with complex behaviour (for example human model or the like).

The unique translation of history-based, parametric and feature models from one native representation into another one is a highly complicated, knowledge-based process. Particularly in case of user-defined features and history-based models, the question of the quality and costs of such translations arises. The other point is, that the standardization of file formats to exchange these kind of models can only slowly follow the actual needs.

The problem is, if there is an actual need to translate parametric or history-based models between different native representations or if there are other methods to submit product data dynamically from one CAx system to another.

Product Data Flow

The product data flow in the virtual product development is no longer sequential. Design activities like concurrent engineering require shared data rather then exchange of a bulk of data. That means, the required product data flow should be a bi-directional stream rather than an exchange of files. It also should build and preserve dynamic links between the original model and the mirrored representation in other systems.

Software Tools Integration

The software tools across the virtual development process are heterogeneous. Dedicated software programmes to solve specific problems are being developed by different suppliers obviously without respect to the whole design process. The application companies deal with this problem by gathering the CAS/CAE and CAM tools around one "master" CAD system. Sometimes there is also a need to use two "master" CAD systems, according to the specific system strengths. For example, in some automotive companies Pro/ENGINEER is used for power train design and CATIA for car body design.

Thus two strategies for dealing with heterogeneous CAx environments can be defined:
* One-system strategy, which means the usage of one CAD system across the whole development process. All additional CAx applications are being integrated under the GUI of this system to increase the ease-of-use of the system. The interoperability between different instances of the system (for different applications) is reached through exchange of native data files.
* Multi-system strategy, which means usage of different systems in the development process according to the system strength, i. e. usage of the „best-of-class" tool for each activity.

3 A Vision of a CAx Component System

A vision of a component CAx system was created to integrate all the views discussed above [1, 2]. It postulates a system consisting of CAx components, i. e. binary software pieces with a common interface to support different design activities. The system can be configured as the process demands and the components of different suppliers can be „mixed" together (plug and play).

In order to implement this vision some pre-requisites first must be fulfilled. A market of flexible, process oriented CAx components of different suppliers is needed. The components should be available through component catalogues. One or more integration platforms for CAx components must be available.

4 CARMEN – A Step Towards the Vision

CARMEN (CAx Remote Model Embedding Network) is an approach to achieve the best interoperability between different CAx systems on the market, not necessarily designed as CAx components as understood in this paper.

The embedding of CAx objects means, that a "mirror" of the original remote object will be created and put into the database of the working system (CATIA in Fig. 1). Since the both systems are linked online, a link between the original object and the mirror object is available during the whole session.

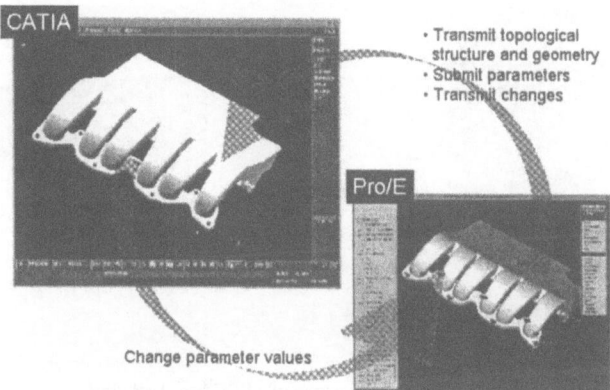

Fig. 1 Pro/ENGINEER object embedded in CATIA

The "mirror" is a regular object in the working system. It's representation is a B-rep, thus exact operations such as cut or distance measures can be performed on it.

The online link between the original object and the mirror allows working with the original object without translation of the parametrics or history. The required operations (for example parameter changes) can be requested and will be performed in the original system. The result will be immediately transmitted to the requesting system.

In this way, native parametrics of the embedded object as well as its full functionality in the original system can be accessed. Also legacy CAx systems can be „componentized" and integrated to achieve maximum process support (see also [4]).

Product Data Channel (PDC)
The Product Data Channel is a runtime environment for CARMEN. It uses STEP as a data model to achieve maximum interoperability across different design activities and CORBA to achieve transparency on heterogeneous CAx systems. The PDC consists of a CAx-object bus (see also [3]) and a set of adapters to specific systems.

The functionality of PDC is as follows:
• Establishing the connection between one or more CAx systems
• Mapping of the topological structures and geometric entities between the connected CAx systems
• Exchanging of the B-rep and preview data
• Maintaining the references between the original objects and the „mirror" in the working system

Based on PDC, flexible process-oriented solutions can be developed. Fig. 2 shows the general architecture of a PDC-based application. Note, that the PDC-AI is the API of the PDC. At present adapters to CAD systems CATIA and Pro/ENGINEER as well as to the human model RAMSIS are available. An adapter to the Unigraphics CAD system will be available soon.

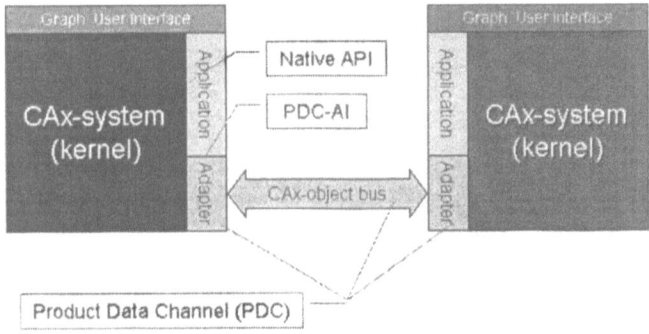

Fig. 2   General architecture of a PDC-based solution

PDC can be used to develop applications in several areas such as Design in Context, Concurrent Engineering, integration of engineering and simulation programmes, video conferencing for CAD (CAD conferencing) and other.

CAD2CAT

CAD2CAT is one of the applications of the PDC. It is used to increase interoperability between CATIA and other CAD systems in the area of industrial concurrent design. It enables remote access to different native CAD models directly from a CATIA session. Viewing, manipulation and interrogation of the embedded models as well as simple parameter changes (e. g. blend radius) can be performed by the CATIA user through an easy-to-use CATIA-like graphical interface. Remote models can be combined with local CATIA models for operations like Digital Mockup (DMU). The current version of CAD2CAT provides a link between Pro/ENGINEER and CATIA. An adapter to Unigraphics CAD system will be available soon.

Fig. 3 shows one of the fundamental functionalities of CAD2CAT – embedding of a partial model. For some applications like DMU it is not necessary to import the whole parts from the remote system. A fast preview allows to pre-select the relevant areas of the model and to transmit only these areas in an exact representation (B-rep).

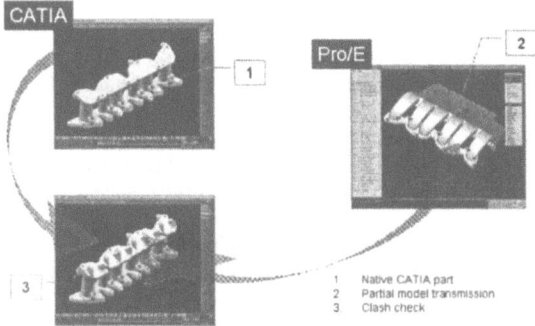

Fig. 3 Embedding a partial model

Once created, a design context can be stored as a regular CATIA session. Because the references to the original parts in the remote system are also stored by CAD2CAT, it is easy to check if the context has changed and to update the differences.

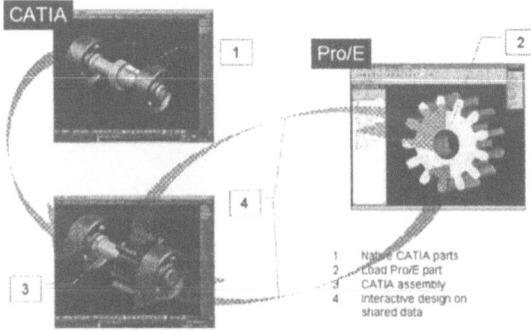

Fig. 4 Clash check in a heterogeneous assembly

Fig. 4 shows an example of another principal CAD2CAT functionality – building a heterogeneous assemblies which consist of models from different CAD systems. It is also possible to work simultaneously on the shared data in this heterogeneous assembly. The differences are updated automatically in both systems to preserve the consistency of the product data.

5 Conclusion

CAx system interoperability is more than product data exchange via neutral file formats like IGES or STEP/PDES. Process aspects (such as product data flow) and integration of software tools in the development process have to be considered. A practicable solution of the interoperability problem should thus represent a *mechanism* rather than a *format* for product data exchange.

CARMEN (CAx Remote Model Embedding Network) is an approach to build up such mechanism and an important step towards the vision of a component CAx system.

6 References

[1] Dankwort C. W., Janocha A. T.: *Von Monolithen zu Komponenten: CAx-Architekturen im Wandel*, in: D. Ruland (hrsg.): Tagungsband CAD '96, Verteilte und intelligente CAD-Systeme, Kaiserslautern, März 1996.
[2] Janocha A. T., Gandyra M.: *Richtungsweisender Einsatz der Komponententechnologie für CAx-Systeme*, in: VDI-Berichte 1357: „Neue Generation von CAD / CAM-Systemen - Erfüllte und enttäuschte Erwartungen", Tagung München, 28./29.10.1997 (anläßlich SYSTEMS '97), S. 87-104.
[3] Iselborn B.: *Entwurf eines CAx-Objektbusses*, in: Anderl R., Encarnacao J. L., Rix J. (Hrsg.): Tagungsband CAD '98 - Tele-CAD, Produktentwicklung in Netzwerken, Fachtagung der Gesellschaft für Informatik, Darmstadt, März 1998, S. 177-187.
[4] Arnold F., Janocha A. T., Kilb T., Swienczek B.: *Rendezvous der Monolithen: Integration heterogener CAx-Systeme*, in: OBJEKTspektrum, Nr. 6, November / Dezember 1999, S. 22-27.

Press Die Design by Parametric Solid Modeling

Hiroyuki Makino
Nissan Motor Co., Kanagawa Japan
h-makino@mail.nissan.co.jp

Abstract: As the result of changes to the lifestyle of the consumer, the automotive market needs are quickly changing. To be competitive in this environment, Nissan has developed Digital Process Innovation. Panel forming simulation and parametric solid modeling are essential in shortening the body press die manufacturing period. This paper discusses how parametric solid CAD can be applied to the shorten the press die design and the full mold Manufacturing period.

1 Shortening the Die Design and Die Manufacturing period

There are two major problems associated with shortening the conventional die design and die manufacturing period.
1).Long die design and full mold manufacturing period
2).Too many die design errors that can interrupt the manufacturing of dies(collision, parts missing,etc)

There area two main causes of the probrems listed above:
1).It takes a long time to draw and read the die design because of complicated die drawings.
2).It takes a long time to make a polystyrene model casting model(polystyrene model for sand casting) by hand.
The difficulty in drawing and reading a conventional die drawing can be seen by looking at Figure 1.In the past, a full model casting model is made by hand from looking at this type of drawing. It takes a great deal of time and there inevitably some differences between the drawing and the casting. This causes big problems for the automated machining process.

To solve these problems, the die design process must shift to a digital process using 3D solid data. The solution involves synchronizing the die design and the full mold manufacturing processes as seen in Figure2. To acieve this, two actions are required:
a). Utilize 3D solid CAD as the primary tool for developing complicated die drawing.
b). Utilize 3D solid data to machine the full mold model.

In order for this process to be accomplished, a very important sonstraint is required. That constraint is that die design changes are not allowed after the design review of the 3D solid data. In return, there are three demands placed on the solid CAD system: Utilize 3D CAM processes(especially full mold modelmanufacturing and automated die stracture machining), reduce die design time, and improve die design quality. Based on these requirements, various software packages were investigated and Cadceus and in-house CAM system called CHP were selected.
CAM System called CHP.

2 The change of Die Design CAD

Starting in 1980, the 2D in-house CAD press die design process was developed. From then, the process advanced to 3D surface CAD and the efficiency of die design and manufacturing increased. However, it is impossible to have a perfect correlation between the plan and sectional views in conventional 2D CAD. Some sections have to be repeatedly checked and do not utilize the data efficiently.

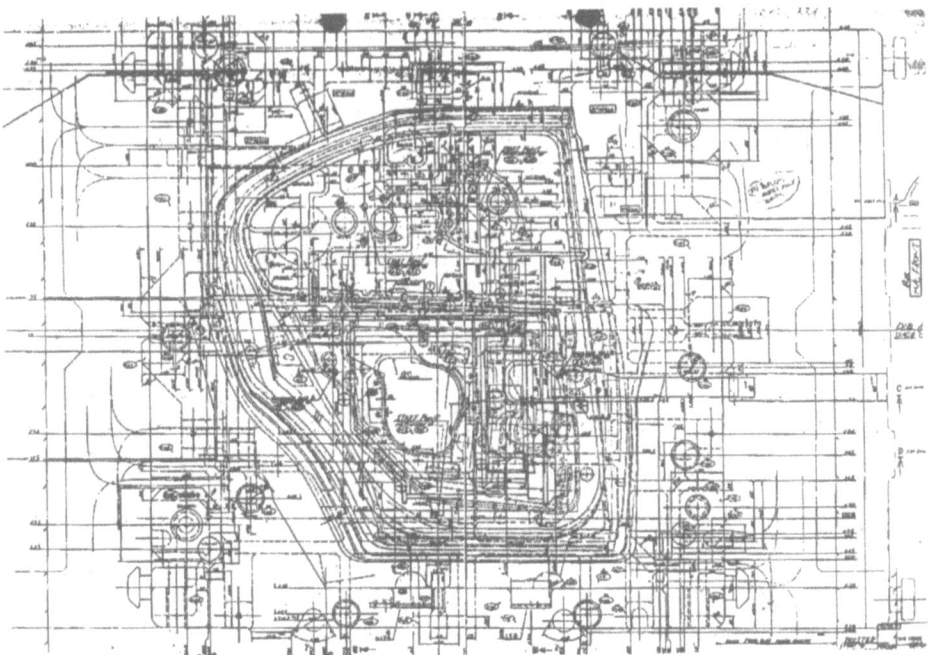

Figure 1 Conventional Die Drawing

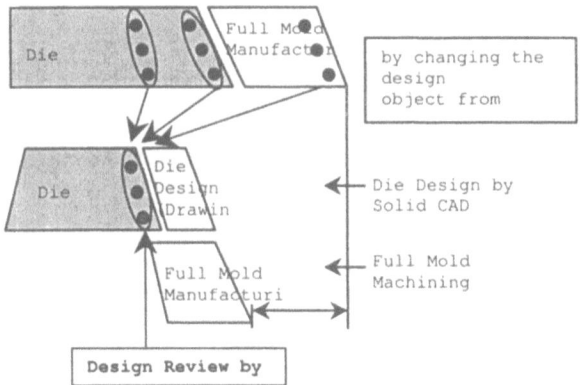

• • Figure 2 Simultaneous Process of Die Design and Full Mold Manufacturing

Recently, by adding 3D solid CAD greater model detail can be added. It also becomes possible to guarantee the design quality by preliminary checking of weights, collisions, etc., and the remaining processes(full mold manufacturing and NC data manufacturing) can utilize the data more effiently via CAM.

Due to the complex and curved surfaces inherent to die design, the size of the 3D solid data usually becomes very large volume. Previously it was diffucult to handle such large quantities of data, but due to rapid advances in computer technology for these past several years it has become more practical over the past few years. Nissan has shortened the die design and manufacturing process by utilizing Cadceus of Nihon Unisys. Co. and an in-house CAM software package. In the next sections, the utilization of CAD data from the die design to make the full mold will be discussed.

3 Die Design Using Parametric Solid Modeling

3.1 Die Design Process

Since it is not efficient to design the die from scratch every time, a standardized die structure pattern is utilized. The pattern is selected based on the part shape and press specification. Please see Figure 3 for an example of standerdized pattern. Each designer can efficiently design by selecting the appropriate standardized stracture and edit it according to the part shape using the parametric function. Later, the die layout CAD data containing detail information about trimming areas, pierce points, etc. can be added to the standardized structure data. Figure.4 shows the scene of applying pierce punch retainer auto layout function. The efficient design became possible using these data and functions.

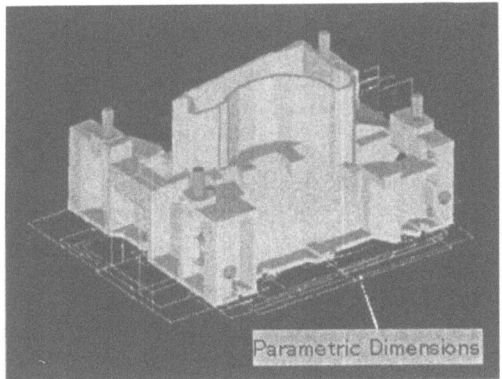
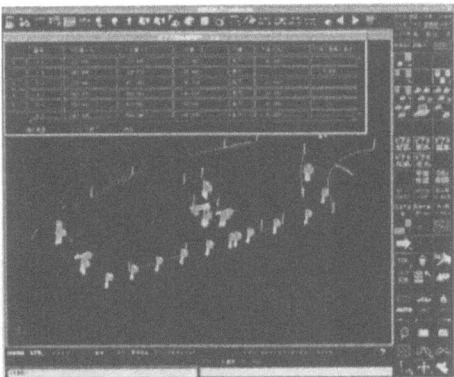

Figure 3 Standardized Die Structure   Figure 4 Pierce Retainer Automatic Layout

3.2 Die Design Review by Solid Data

In order to accomplish simultaneously the die design and the full mold manufacturing, the die design review must take place by viewing the 3D solid data. Various items can be checked by utilizing the 3D solid data. For example, Figure 5 shows a collision check within the die. The Clipping Function (the process of viewing continuous cross-sections across the die structure similar to a medical CT scan) is utilized in the design process and then an automated collision check is done upon design completion. All of the cross sections could not be checked using the conventional 2D drawing, but now this can be done easily using the Clipping Function on the solid data. Additionally, for transfer presses the transfer tooling(fingers, rails, etc.) and panel motion curves and collision can be checked using the Kinematic Function as seen in Figure 6.
Figure 7 shows the Die Rigidity Check using Analysys Module. This example is checking the deflection of a section of the blank holder for a body side outer panel. The results of this check will be incorporated into the design and machining (via NC data) of the die face clearance.

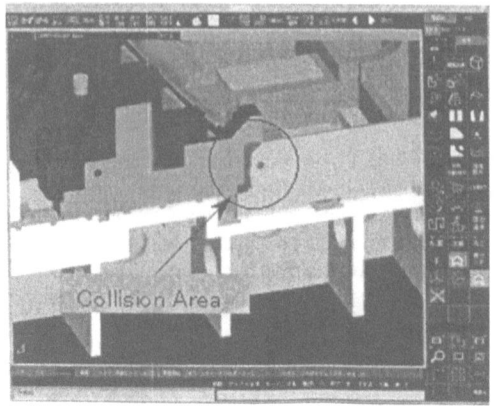

Figure 5 Collision Check inside of Die

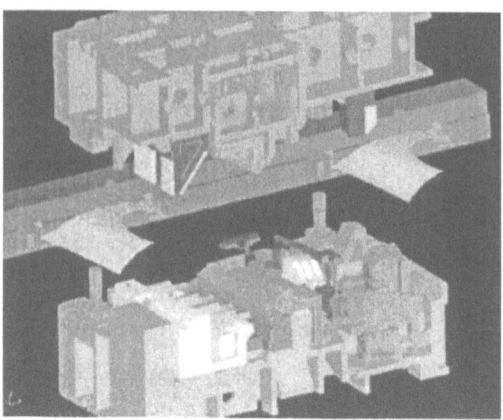

Figure 6 Collision Check of Die and Transfer Tooling and Panel

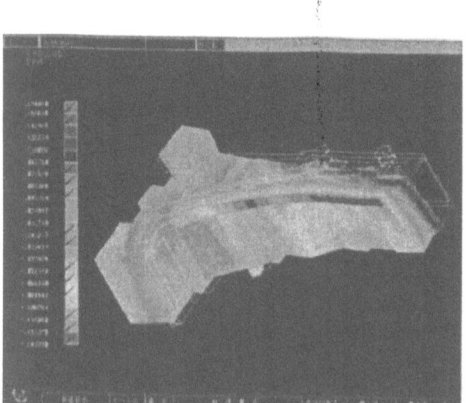

Figure 7 Die Rigidity Check

4 Utilization for CAM Process

The manufacturing process of the die will begin as soon as the design review by solid data has been completed. Next, the full mold pattern will be machined from this solid data.High speed machining can be accomplished with the use of special developed cutting tools as seen in Figure 8. This process not only reduces costs and shortens the time required to produce dies, it also eliminates mistakes whitch can occur when the polystyrene mold is made by hand. Futhermore, casting allowances are automatically added for CAM features like holes and flat surfaces whitch have been added to the solid data. NC data can efficiently analyze these type CAM features and therefore reduce mistakes. (Figure 9)

5 Simplified Drawing

When shifting from using 2D to using 3D data, the requirements of the die drawings change. The full model mold and the NC data manufacturing do not require the detailed dimensioning seen in conventional die drawings. Figure 10 shows an example of simplified drawing. Die design time can be reduced since the dimensioning detail is reduced.

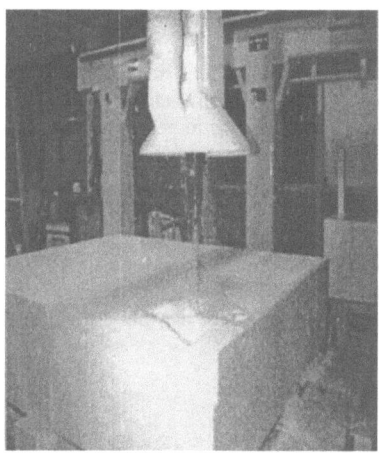

Figure 8 Full Mold Model Manufacturing by Rapid machining

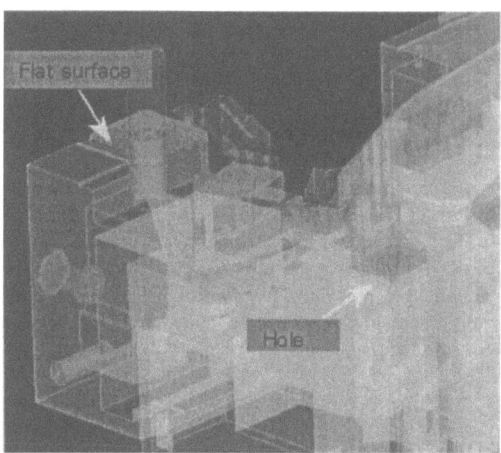

Figure 9 CAM feature for Automated Die Structure Machining

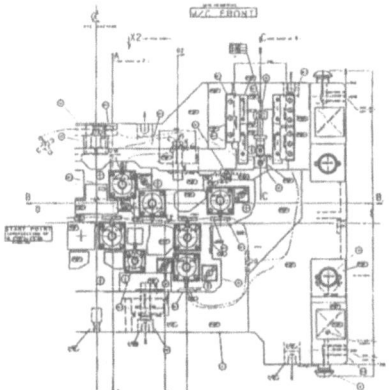

Figure 10 Simplified Drawing

6 The Effect of Solid CAD and Future Subjects

The time required to complete a die design and full mold model can be reduced by 30 percent by utilizing solid CAD data. By utilizing the digital process, die components will be made exactly as the data itself. This results in eliminating the time required to correct differences of each parts whitch occur within the 2D process. To obtain the benefits of reduced die design and manufacturing tme, the following two items need to occur:
*Eliminate losses in the die design and manufacturing process by utilizing die drawing-less activity.

*Eliminate the need of fitting adjustment during die manufacturing by utilizing the digital data.

To gain further benefits, more is required to the software. The software needs to be improved in the area of modeling function in order to more easily and precisely create complicated die shapes. Also if the cost of software and hardware is sufficiently reduced, then Personal computers and Viewer can be used to view and manipulate the die data thoroughout the die design and manufacturing process. And as the network environment improves, the large volume of data can be easily shared with not only the manufacturing sections and some cooperating makers, but also the vehicle assembly press sections.

There are new challenges, to the die design and manufacturing sections. The first area involves the reorganization of engineering and skill. For example, a new engineer typically starts by drawing die design, but due to the shorten design cycle, there is no longer adequate time to permit training. Thus the education method has to be improved. The same concern applies to the manufacturing section. At first they typically start by reading and checking the die drawing, but now they must improve their skills and analyze 3D solid data.

The second area involves the proper selection and usage of CAE application tools and the hardware. It will be increasingly important to make the appropriate choice. Although there are still some challenges, the die development process has already seen a big improvement through the digital process innovation over the conventional drawing method. This is no doubt the future trend of die manufacturing.

Numerical Simulation in Product Development

Günter Müller

CAD-FEM GmbH, Grafing, Munich
Gmueller@cadfem.de

1 Introduction

This paper will focus on simulation of products and to some extend to manufactural processes. It will not cover simulations done for plant design, facilities, logistics, robotics or material flow.

Product simulation tools allow to simulate the behavior of products before they are actually built. Basically two methods are used: Multi Body Systems and Finite Element Methods. Multi Body Systems idealize the real world system by a number of simple rigid and – limited – flexible parts which are connected by various types of hinges. This approach allows to handle complete structures and movements over a long time period. Vehicle dynamics is a perfect application in which the forces acting on a driving car, truck or railway car is investigated. The results are time histories of displacements, velocities, accelerations and forces.

The Finite Element Method (FEM) approximates the real systems in much more detail by a large number of one-, two- and three-dimensional elements. FEM delivers besides displacements and forces also stresses and strains at any point of the structure. Thus this method offers a detailed approximation of the behavior of a structure under given boundary conditions and loads.

Both methods are complementary. Quite often Multi-Body-Systems are used to determine the forces acting on a part which are then used in a following FEM analysis. In this paper we will focus only on the Finite Element Method.

1 Finite Element Method

The physical behavior of structures or processes is defined by differential equations. Dependent on the geometric dimension, the time dependency and the physical phenomena to be considered various equations exist. Examples are statics and dynamics, heat transfer, electromagnetics, fluid flow, acoustics, and coupled fields like piezoelectrics.

When the unknown function in the differential equations is determined the problem is solved. For example for a statically loaded structure with one dominant dimension (length) the so called beam theory applies which is defined by a differential equation of the 4.th order. The unknown function to be determined is the displacement. From the displacement stresses and strains can be derived.

Solutions of the differential equations can be found in two ways: first in solving the differential equation directly, second in solving the corresponding variational problem. In terms of mechanics the latter means that the equilibrium state (defined by the differential equation) is approached by minimizing the potential energy (minimization of the variational function).

For both procedures an analytical ("exact") or numerical (approximate) solution technique can be chosen. Analytical methods are typically limited to simple geometries and boundary conditions. Practical real world problems are usually too complicated for analytical methods. Only with numerical methods such problems can be solved. Numerical methods approximate the exact solutions by a series of functions which are products of selected shape functions and unknown parameters.

Today FEM is the most powerful numerical method. The specific approach of FEM lies in the selection of the shape functions and the parameters. The shape functions are simple functions which only span a sub-domain – called element – of the total domain. The parameters are mechanical values at the nodes which connect the elements. That means that the function which governs the solution of the differential equation is approximated by simple functions over a number of elements. FEM follows the variational approach. By introducing the approximate functions into the variational minimization procedure a system of algebraic equation is generated from which the unknown parameters can be determined. Thus FEM transforms the problem of solving a differential equation into the problem of solving a corresponding system of algebraic equations. Examples of algebraic equations for three disciplines are noted in fig. 1.

The main advantage of FEM is its general applicability. For different disciplines the unknown parameters are different, however the overall solution procedure is similar. For nonlinear materials or large deformations the algebraic equation are nonlinear and have to be solved repeatedly and for time dependent problems several analyses over the time range have to be done. Depending on the size and dimension (1., 2-, 3-dimensional) of the geometry and the requested accuracy the algebraic equation may become very large. Today a typical size might have 100.000 of unknowns but also problems with several millions of unknowns are solved within hours.

For more details about the theory the reader is referred to[1].

Statics, Dynamics:

$$M\,v + C\,v + Kv = F(t)$$

v=displacement,velocity,acceleration. vectors

K = stiffness matrix

F = force (t = time) vector

M = mass matrix

D = damping matrix

Heat Transfer:

$$CT + HT = Q(t)$$

T = temperature vector

H = conduction matrix

C = capacity matrix

Q = heat flux vector

Electromagnetics:

$$C\,A + KA = F(t)$$

A = potential vector

F = current density vector

K = magnetic permeability matrix

C = electrical conduction matrix

Fig. 1 Analysis Types

2 History

The method was developed in the early sixties. The roots of the method are various. John Argyris can be seen as one of the founders because he wrote down static and dynamics equations in matrix notation. Clough and colleagues divided a membrane into sub-domains and put them together under compatibility and equilibrium equations. It was Clough who coined the name "Finite Element Analysis" in 1956 [2]. Thereafter many researchers were working in this field and it was found that the method can be seen as a piecewise approximate solution of a variational method which corresponds in mechanics to the principle of the minimum of potential energy. An idea which was already published by Courant in 1943.

In 1960 first computer programs became available. Famous became STRUDL and ASKA. They were mainly used in large data centers of aerospace institutions. In 1970 the first commercial products came on the market among them MSC/NASTRAN, ANSYS, MARC, ABAQUS and some others. The programs at that time ran only on large computers, so called mainframes. Towards the end of that decade the first minicomputer like the famous VAX 780 could be used for FEM analysis. A second wave of programs was introduced around 1980 with programs like ADINA, COSMOS, ALGOR, to name a few. At that time some of the programs were ported on personal computers. CAD-programs like COMPUTERVISION, I-DEAS offered modules for FEM, however, they were quite limited and the acceptance was low because it was still too complicated to run an analysis - only recently this has changed. Towards the end of the 80-ties programs based on explicit solution techniques like LS-DYNA3D (now LS-DYNA) and PAM-CRASH became available for crash and metalforming analyses. In the 90-ties fluid flow became more popular and codes like STAR-CD and FLUENT became established.

3 Benefits and Concerns

The main benefits of the application of the FEM are: reduction of development time, reduction of cost, and saving of resources. It allows also the design of innovative products with better quality and it allows to quickly comply with new regulations.
Manufacturing cost can be considerably reduced if more time is spent up-front. Changes which have to be done at the end of the development or even at begin of manufacturing are very expensive. Higher quality and innovative products are important to compete with products made in low-cost countries. Many new regulations are defined by the European Community and stricter laws for product liability have to be followed immediately. Simulation starting from the very beginning of product development can help to meet all these requirements.
There is not much material available about the usage of FEM in industry. One study was done at the Technical University of Munich [3] The study says that 63% of the surveyed companies use simulation, 48% of smaller companies and 83% of larger companies. The actual numbers are estimated to be lower because it is assumed that those companies who have not used simulation at all have not responded. With 82% automotive is the clear leader followed by machinery (57%) and electronics (45%). The main motivation to use FEM is to improve the product design (75%).15% have done the analysis because it was required by the client. 80% of the users feel that they had benefits. The applications are mostly in mechanics (75%), only 20% do thermal or fluid flow and less than 5% perform electromagnetic analyses so far. Most applications are done in product development, less than 10% have applied FEM in process simulation.
Obstacles to implement the method are: investment in hardware and software (60%), additional effort (time) (30%) no permanent need (30%) and lack of qualified employees (20%). One other important obstacle is also the time needed to generate the FEM-model. This problem will be reduced in the future because more and more the already generated geometry from a 3d-CAD-program will be used. Today exists still the problem that the CAD-geometry is too detailed and mathematically it is often not correct. For example there might be a gap between two lines and therefore a FEM mesh not connected. More development effort is still required to get a valid geometry for FEM analysis on the basis of CAD- geometry.

4 Applications Today

At the beginning FEM was mainly applied in the aerospace, power plant, civil engineering and offshore industry. One of the major users today is the automotive industry. One very extensive usage is in crashworthiness studies which are very expensive if done in experiments. According to General Motors (1997) one test costs US $ 750.000. FEM reduces the number of necessary tests dramatically. BMW is doing 91 crash-simulations and thus is able to reduce the number of real tests to only 2. The 91 computer tests cost less than the 2 real tests [4]. Crashworthiness studies started in 1985. The models at that time consisted of about 5.000 elements and it took the most powerful computer 10 hours to solve the problem. Today the car models consist of more than 250.000 elements and the computer time is still almost the same if not less. There are many different crash-simulations to be performed: front (central and offset), side and rear. So far most of the tests are done against a rigid wall. Now also compatibility studies of large cars impacting small cars are done. Smaller cars have to be designed stiffer and large cars more flexible to give both drivers a chance to survive.

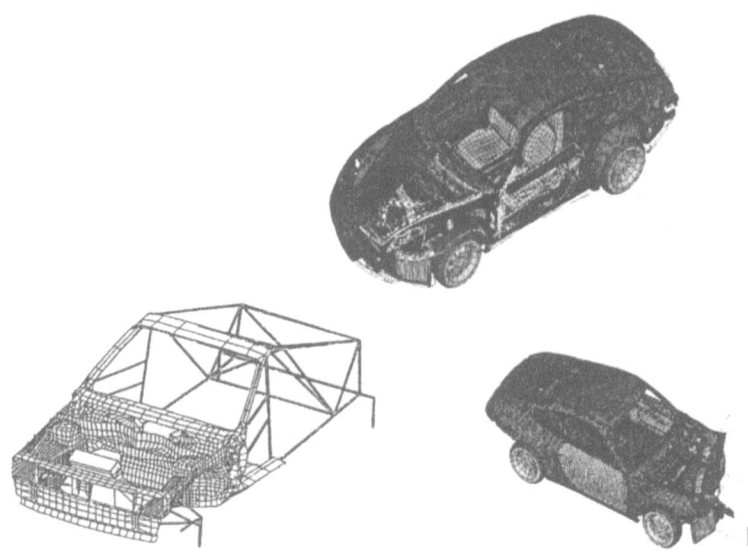

Fig. 2 Crashworthiness Studies 1987 and 1998   Courtesy of Porsche AG

Today crash analysis is also done for busses, trucks, railways, ships and even airplanes. FEM has found a lot of credibility so that a new regulation of the European Community even allows numerical roll-over-test for busses. Trucks to be exported to Sweden have to fulfil the so called Sweden test which is basically an impact on the top, front and rear of the drivers cabin.

The method has spread into all other industry branches. In electronics the devices become smaller and smaller and more heat has to be dissipated. The devices are so small that even different physical phenomena interact. This allows to design new types of sensors. Multfield analyses have to be performed which include displacement, electrostatic, electromagnetic and temperature fields and even fluid flow. Machinery is another large application area, however, in tool machinery the applications are still rather limited.

An example where FEM was used with great success from the very beginning all over the design process till the final design is the project of the European Southern Observatory. The task was to build the largest telescope with a mirror of 8 m diameter. At the beginning nobody had an idea how the construction would finally look like. Various designs were investigated and analysis results helped a lot to make the right decisions. All different analyses like static, frequency response, earthquake, heat transfer and fluid flow analyses had to be performed.

Fig. 3 Very Large Telescope - Courtesy of European Southern Observatory

Also in science FEM is used a lot: to predict earthquakes, to simulate the flow of blood in the arteries, to investigate the statics of plants or the bones or teethes of animals to name just a few applications.
Process simulation in general is a topic which is just at the beginning. Applications can be mainly found in metalforming and forging.

5 Trends and Requirements

CAE will grow steadily. 3d-CAD-programs will replace 2d-systems.
FEM will be used by the designers. The acceptance problem seems to be overcome. Easy to use, robust and reliable FEM-Modules are now available for example in Pro/ENGINEER and CATIA. DesignSpace from ANSYS, Inc. is another product which is extremely easy to use and which is tightly coupled to the main CAD-systems. With these products the design engineer is able to solve simple standard linear static and thermal analyses. However, it must be emphasized that a thorough training is mandatory to assure that the analyses are done correctly. The user needs a basic understanding of the theory and the solution procedure and he must know the limits of the method. The traditional analyst will not become obsolete. He will act as a coach to the design engineer and he will spend his time on more complex problems which will increase.
Today analysis and CAD are still more or less separated. In the future a tighter link to PDM-systems is required to get all data from one source. In addition interface programs are needed which fix the flaws of geometric data and allow to defeature the geometry which stems from CAD-systems. Another approach which might be followed is to start from a simplified geometry, do all the necessary analyses and optimizations Only after the simulation has proved that the product fulfills all requirements for its later use a detailed CAD model is generated
The number of programs will decrease. This is a trend which already can be seen in CAD were only a handful of systems are left over.
Customers will stay with the established programs like MSC/NASTRAN, ANSYS, ABAQUS or LS-DYNA, but there will be a need for additional features. That means that customized vertical applications to the established programs have to be developed.
Displacement and stresses as result will not be sufficient. More emphasis will be given to the prediction of life cycle. This means, fatigue analysis and probabilistic design will become main topics.
Optimization methods are available today, however, the implementation is not done in such a way that it can be used easily in the design environment. Also the number of parameters and production criteria are still limited. Shape optimization and topology optimization are still done separately. In future they will be merged.

New analysis methods are emerging. For example the particle method which is not based on the differential equations of the problem, but approaches the problem using interacting particles. This approach omits the still tedious mesh generation process, because no mesh is needed.

The size of problems is getting bigger. Models with millions of unknowns will become not unusual. This and the trend to nonlinear and multidisciplinary analyses will need parallel shared and distributed processing. This is already available today, but will become more important in the future.

Implicit and explicit algorithms will be merged to make use of the specific advantages of both methods.

In crashworthiness the vehicle to vehicle crash will be studied more extensively and also pedestrian safety becomes an issue. By 2004 a pedestrian must survive an impact with a car running at 40 km/h. This is not an easy task because on one side a car should be stiff enough to withstand a crash against a rigid wall on the other hand it must be flexible enough to limit the impact force on a smaller car or a pedestrian.

Fluid flow is not yet as established as mechanical analysis. There is a large potential of applications in that area. Programs like STAR-CD, CFX, FLUENT and PowerFLOW (based on particle method) become more and more popular. Fluid-structure interaction problems will become most important.

One large application area will become process simulation. The industry is very creative in inventing new processes to reduce the cost of manufacturing. One important process is metalforming which includes deep drawing, hydroforming and forging. Solutions are already available, but more development is needed to improve speed, accuracy and ease of use. Realistic friction modeling and springback analysis need more investigation.

Other processes to be simulated are casting, powder metallurgy, welding, blow molding, thermoforming of plastics. An example of thermoforming is shown in fig. 4. The simulation allows to determine the distortion of a picture on the original form to get an undistorted picture on the final shape.

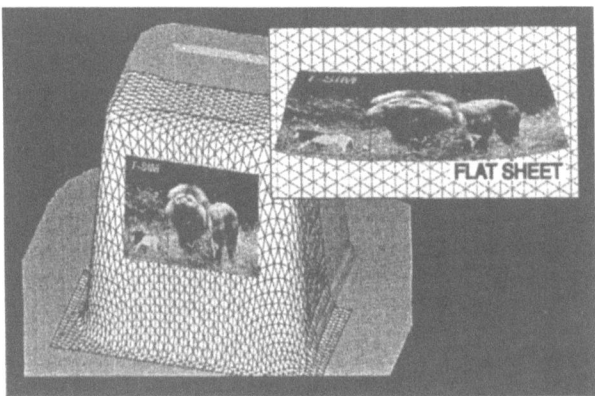

Fig. 4 Thermoforming Simulation

A very new application of FEM is in simulation of the complete process of spraying of a car body. This application helps to save high cost and much time. To simulate the complete process a number of simulation tasks have to be performed. The drying process is just one. During a given time the temperature of the body surface must be in a given range to assure a high quality of the paint. Such a simulation can be done by a simple heat transfer by using suitable heat convection values. A coupled fluid flow / heat transfer analysis would be most appropriate, however, such an analysis today requires too much time and resources, at least today. This might be different in the near future.

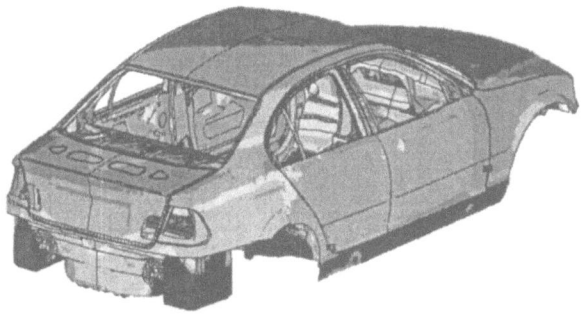

Fig. 5 Drying of the Paint of a Car Body   Courtesy of BMW AG

In food industry lies also a large potential. So far only a few applications are done, like the simulation of salt concentration over time in parmesan cheese or the drying of pasta, which is a multifield analysis including unknowns like pressure, humidity and temperature.

6 Outlook

The ultimate goal is to have one digital model on which all kind of necessary studies can be done. An optimization strategy would take care that all requirements are met and still an optimum in weight or cost is achieved.

The car manufacturers are moving into this direction. The aim is to develop a digital car which allows to investigate the behavior of the new design before it is built. A step in this direction is done in the software eta/VPG which uses one model of a car for a test on a virtual proving ground. This analysis can deliver data for fatigue, noise and vibration on the same time. Today a number of analyses with different models and programs are done. One unique model for all type of analyses, an acceptable amount of computer time and a bidirectional link to CAD and PDM is the ultimate goal in simulation.

The job environment will also change. Internet will allow engineers located at different sites to collaborate and work simultaneously on the same model.. Engineers from various companies in different countries can team up for a project and after finishing this project can take part in another project.

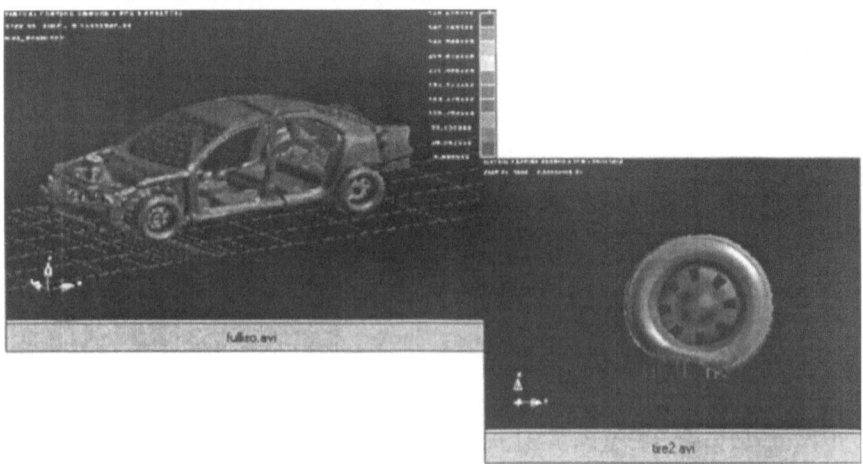

Fig. 6 Virtual Prooving Ground

7   References

[1] K.-J. Bathe. Finite Element Procedures in Engineering Analysis. *Prentice-Hall*, 1982
[2] R. W. Clough. The Finite Element Method in Plane Stress Analysis. Proceedings. ASCE Conference. On Electronic Computation, Pittsburgh, PA, Sept. 1960
[3] G. Reinhard, K. Feldmann. Simulation – Schlüsseltechnologie der Zukunft? Stand und Perspektiven. Herbert Utz Verlag, München, 1997.
[4] BMW, Crash am Computer, Automobil-Entwicklung, September 1999

Parametrics: Present and Future

Akihiko Ohtaka

Nihon Unisys, Ltd.,1-1-1 Toyosu, koto-ku, Tokyo 135-8560, Japan
Akihiko.Ohtaka@unisys.co.jp

Abstract: The present day parametric modeling technology is playing an important role for the efficient change of product shape, for representing some part of design constraints, and for the improvement of re-usability of existing shape models. Though it provides a powerful shape modeling environment, there remains future issues which require further improvement of the technology in order to realize a real design tool. The present paper first reviews current parametric technology from the view points of underlying technology, effective applications and data structure. Then, the author proposes some ideas for making it more design tool oriented.

1 Introduction

The power of present day parametric modeling technology used in mechanical CAD/CAM area is gradually recognized by end users and it may have stronger impact than the appearance of solid modeling in the past. The major reasons are that the parametric modeling technology has a possibility to explicitly represent design process or design intent which no other shape modeling technology allows, and combination with a variety of domain specific applications such as die design, or more generally, with CAE and CAM could drastically improve product manufacturing.
Standardization of parametrics is continued in ISO TC184/SC4 within the context of STEP with which the author is concerned.
This paper first reviews constituent technology of current parametics in chapter-2. Then future issues are discussed in chapter-3. The author proposes some technical improvements of parametrics in order to make it more suited as a design support tool in chapter-4.

2 Review of Current Parametrics

2.1 Target of Parametric Behavior

In current parametrics, the target which is allowed parametric representation is limited to product shape.

2.2 Constituent Technology

Constituent technologies of current parametrics are;
1) 2D parametric representation of product shape by the use of geometric constraints.
All the commercial systems support free hand sketch function where the system navigates users during modeling operations in order to let users know geometric constraints automatically accumulated
2) 3D parametric positioning between parts by the use of geometric constraints
3) Operations history based parametric representation of product shape
4) Algebraic relation among variables.
1), and 2) have sound background mathematics and CAD system dependence of representation specification is quite small. These technologies are non procedural which implies that constraint solvers do not care about definition sequence of constraints.

3) on the contrary, memorizes all the operations history and related information for later robust reconstruction when parameters are changed. Since operations history contains bare modeling operation(CAD command) of which specification is different system by system, CAD system dependence is considerable. This technology is procedural since solving systems replay operations history with given order.

4) is used in all the technologies 1), 2) and 3) for representing constraints among variables which are beyond the scope of geometric constraint representation. CAD system dependence here is quite small. Some solver applies procedural solving method but others apply non procedural solving method.

In ISO/STEP parametrics, there are one NWI(New Work Item) which treats 1), 2) and 4), and one PWI(Preliminary Work Item) which treats 3).

As for the relation of these technologies, 3) plays a central role in all the major CAD systems where 1), 2) and 4) are combined with it when necessary.

2.3 Possible Applications

The use of current parametrics is centered to detail product shape design phase. Typical possible applications are;

(1) Parametric part shape modeling
Part shape modeling operations are performed in parametric mode where the system automatically accumulates necessary information for the later reconstruction.

3) above plays a central role, but 1), 2), and 4) are also used as necessary. Typical example is creation of a protrusion on top of existing part model. The user will perform the job with the following steps;

   a. Create 2D profile of the protrusion by the use of sketch function(use of 1) above)

   b. Rotate it 360 degrees for completing 3D primitive(use of 3))

   c. Perform a set operation for merging the protrusion on top of existing part model(use of 3))

   d. If constraint among variables(mainly, dimensions) are necessary, the user will define pertaining relations(use of 4))

After the completion of the part shape, various shape changes are made available. Operations history accumulated can be effectively used if any shape change becomes necessary. The user directly specifies the target shape which should be changed. Then, the system shows its defining parameters. After the user changes parameters as necessary, he requests the system to reconstruct. Then, the system automatically reconstructs the part shape according to the given parameter changes guaranteeing whole topological and geometric consistency. Here, sophistication of the reconstruction algorithm becomes a problem. For example, the shape modeling of a cylinder head of a car engine could cost thousands of modeling operations. When requested parameter change is for an operation which is located in the near end of operation history, there is no need to reconstruct whole part shape. Some commercial system supports minimum reconstruction function which only recalculate least necessary operations. Most commercial systems provide a variety of operations history editing functions. Examples are insertion of a new operation, replacement with a new operation, go back to arbitrary past operation, exchange of two operations, and inactivation of operations. Inactivation of operations is effectively used to idealize a part model shape so that it is suitable for FEM analysis or rough milling by inactivating operations which create small holes or small fillets.

(2) Parametric assembly modeling
Only one difference with parametric part shape modeling is that the target of reconstruction is not shape element but transformation matrix between parts. Except for this, all which is written in (1) also applies to parametric assembly modeling.

(3) Collaborative design support
Collaborative design support which is one of the major future issues could be eased by the use of parametrics. One of the major issues there is how to automatically reflect related engineers design change. Reference change driven automatic reconstruction of a part shape which is available with current parametrics can support this requirements as shown in Figure-1.

(4) Association with CAE and CAM
Idealization of product shape for use in CAE and CAM as mentioned above is already in practice.Parametric product shape change based CAE has a strong possibility to promote optimal design.

Association with NC is also hopeful for realizing real integration of CAD and CAM.

2.4 Representation of operations history

How to represent operations history is an important issue for making history based patrametrics robust or making it more technically sound. The followings show main points of parametric data model implemented in a commercial system CADCEUS[1]. Figure2 shows basic data structure.

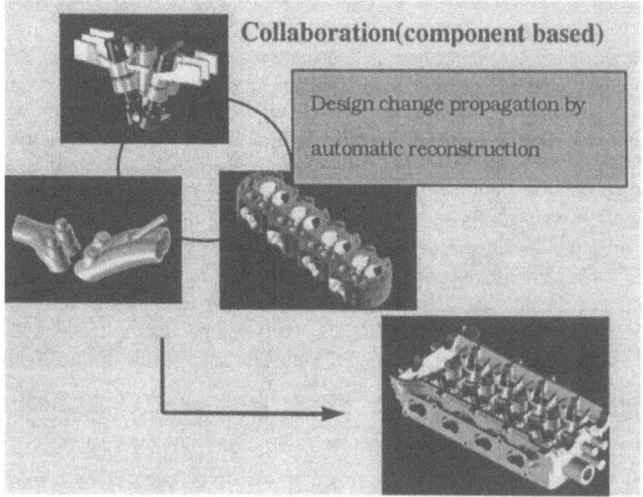

Figure1: Use of parametric reconstruction in collaborative Design

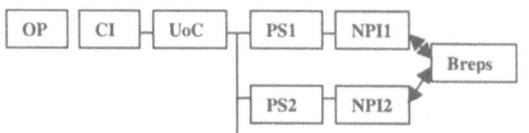

Figure-2: Basic Parametric Data Set

OP which stands for operation corresponds to one modeling operation such as swept surface creation, set operation, copy, delete, etc. The following status information is memorized with an operation;
1)operation sequence number
2)if it is inactivated or not
3)if it is waiting reconstruction or not
4)if pertaining parameters are changed or not
CI which stands for command instance memorizes all input information for that operation, and additional information necessary for later reconstruction. Information memorized are;
1)input parameters
Input parameters include numerical values and reference geometry. If a fillet command requires identification of rounding edge and fillet radius, rounding edge id together with related surface ids are memorized. The reason why surface ids had better be memorized is that surface id based reference maintenance is advantageous from the robustness point of view. In the same context, point id plus related surface ids are memorized in point reference case. In case a sketch which describes sectional

---

[1] 1: CADCEUS is the integrated CAD/CAM/CAE/CG system developed by Nihon Unisys. CADCEUS is adopted as the base of Toyota's 'TOGO' system, and has many users especially in automobile industry.

geometry is used in a boss creation command, the sketch id should also be memorized. Here is a discussion how to treat a sketch, which is comprised of geometric constraints, in history based circumstances. The author's experience shows that one sketch, even if it contains a number of geometry, can be treated as one operation.

2)additional information

The following information which relates to modeling circumstances should also be memorized for robust reconstruction.

  Current dimension (2 or 3)
  Current coordinate system
  Current group
  Current layer
  Current instance

UoC which stands for unit of creation summarizes several units of shapes(features) created by one operation. Typical example is a patterned feature creation command where two or more features are created by one operation.

PS which stands for pseudo structure is prepared for summarizing a unit of shape created by one operation(command). Contents of PS differ depending on the nature of commands. In solid or surface creation commands, PS shows created solid or surface data. Commands which make some changes to existing shape such as 'move' or 'delete', do not have corresponding PS. In commands which change existing shape such as set operation, newly created curves by the operation become constituents of PS, but whole solid or trimmed surface are not regarded as PS.

PS preserves the original data structure just after the creation of a shape by an operation. PS is also prepared for enabling structured naming mechanism. Relation between parametric model and whole Brep model is maintained via geometric entity under PS. PS could also serve a role of identifying form features.

NPI which stands for name preserving information supports name preserving mechanism and maintenance of semantics of data. As a typical example, the followings show the contents of NPI for swept area operation.

Suppose as a typical case swept body creation operation where the area to be swept is represented by a sketch. NPI information becomes as shown in Figure-3.

Even in the case some part of the original sketch geometry is changed, unchanged portion can be maintained with the same surface ids as shown in Figure-4.

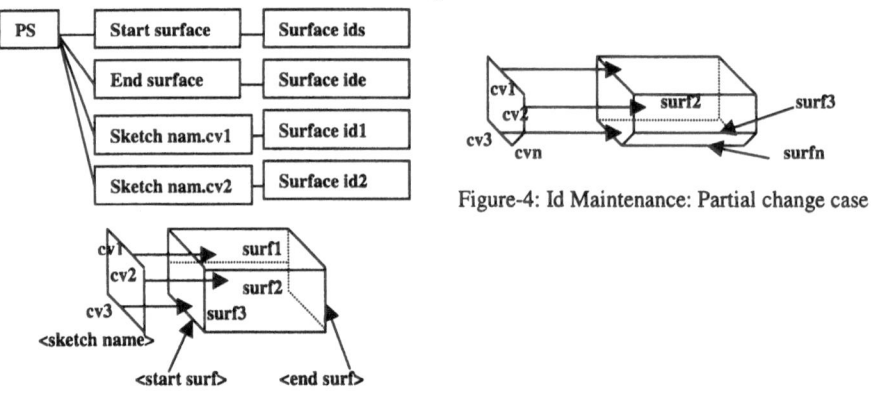

Figure-4: Id Maintenance: Partial change case

Figure-3: NPI for sweep operation

In case some edge of the swept body is later rounded, edges and surfaces are divided but it is similarly possible to maintain the same edge ids and surface ids for unchanged portion.

In this id maintenance mechanism, original sketch should not be replaced but should be modified.

Of course, fillet operation is dependent on the original swept body operation since that fillet is meaningless if the swept body does not exist. This kind of operation dependence is maintained and strongly affects reconstruction

3 Future Issues

As mentioned in chapter-2, most future issues of parametrics exist in history based parametrics. Though it is quite powerful even now and it has hopeful future applications, resolution of issues mentioned below will greatly improve reliability and applicability of the technology. Major issue the author understands are as follows;

(1) How to improve human readability of operations history
Operations history currently implemented in commercial systems memorizes all command used during modeling in given order, and some other information which is necessary for later reconstruction. First problem is that designers intention is nowhere described and only commands used are memorized.

There are frequent claims from end users that current operations history can not be shared even in the same design department since unexpected results frequently occur when a designer perform design using an operations history originated by other designer. Second problem is its volume. Mechanical part modeling like a cylinder head modeling could consists of more than two thousands operations which is difficult to read or understand. Basic reason of this enormous volume is that current one CAD command does not correspond to one functional design unit. It is usually the case that one functional design unit is represented by successive two or more commands.

(2) How to eliminate CAD system dependency
Major constituents of operations history is CAD command used during modeling of which specification is CAD system dependent. This issue is critical in standardization activity.
Other information memorized for later robust reconstruction or minimum reconstruction are also different system by system.

(3) How to represent meaningful design process and design intention
Designers never fail to contemplate a strategy how to reach final target shape based on his design experiences and knowledge about the CAD system in use. Operations history is its result and his thinking is not explicitly represented there.

There should be some mechanism to capture or extract essential design process and critical design intention of each process.

(4) Target of parametric behavior is limited to product shape modeling commands which is a major bottleneck for making it a design tool oriented technology.
There are two discussion points here. The first is target shape model. Most of commercial systems allow parametric modeling only for solid model. Major reason could be that solid models are advantageous from robust reconstruction point of view since they have enough information compared with surface models or wireframe models. But, sophistication of reconstruction algorithm has a possibility to extend parametric behavior to surface model, wireframe model or even their hybrids which leads to wider applications.

The second is target commands which are allowed parametric behavior is limited to commands which make changes to product shape. This is the reason why current parametrics is called 'Shape Parametrics'. But, real shape design is basically the repetition of shape change and its evaluation.

Any designer does not proceed his design until he is satisfied that current shape change does satisfy his design intention. This evaluation is done with a variety of analysis commands such gap analysis, surface quality analysis, stress analysis, etc.

(5) Persistent naming issue
The most critical issue with which all the commercial system vendors are struggling is how to realize persistent naming. During modeling, designers freely refer existing shape element for defining new shape element. The system memorizes dependence relation between these two elements. At the time of reconstruction, memorized reference id could become undefined caused by the change of original shape element such as division, deletion etc. In order to avoid such unsuccessful reconstruction, the system should have a mechanism which always guarantees well defined reference. In other words, maintenance of robust reference relation is the critical factor for realizing robust reconstruction.

4 Some Proposals for Improving Parametrics

In order to resolve issues mentioned in the previous chapter, drastic innovation of CAD system towards functional modeling, further research and implementation of the representation of design process and design intention are the necessity. Those proposed below are aimed at pragmatic improvements of current parametrics until above mentioned challenges result in clear answer.

Structured representation of operations history
In order to improve human readability and re-usability of operations history, the author proposes structuring of operations history with adequate user definable attributes for describing critical design intention. The followings show an example in cylinder head modeling case.

```
OP class(process name: whole design, assertions:[1:?]of assertion)
   OP subclass(process name: exhaust port, assertions:[1:?] of assertion)
      OP1
      OP2
      •
      •

   OP subclass(process name: water jacket, assertions:[1.?] of assertion)
      OP1'
      OP2'
      •
      •

   OP subclass(process name: cylinder head, assertions[1:?] of assertion)
      •
      •

END_OP
```

Figure-5: Structured Operations History with Attributes

In Figure-5, an OP class corresponds to one design process which designers recognize. OPs below an OP class show actual modeling operations for realizing target design process satisfying assertions stated in assertions field.
Assertion is prepared to explicitly declare conditions which validate that OP class. Free algebraic relation such as
   Distance (surf1,surf2) < 3
   L1 = 2*L2 + L3
can be described there.
In case description with this types of mathematical representation is not possible, the designer may state his intention with natural language as comments. Basic understanding here is that operation itself could be different designer by designer depending on their design experiences and knowledge about the CAD system in use, but assertions which validate that design should be shared.

Design evaluation
The designer evaluates his design at necessary points if his design is valid, or his design satisfies various constraint conditions. There are evaluations which require heavy simulation for judging stress distribution, warping, temperature distribution, material filling analysis, etc.
But, most evaluations are more simple which require evaluation based on prescribed check sheet comprised of design criteria. In order to include design evaluation into parametric world, the followings are necessary.
1) Allow analysis commands in operations history where the most focused value the designer cared among analysis results is made clear.
2) Allow user definable design check sheet in operations history.

If these two requirements: structured representation of operations history and inclusion of design evaluation are realized, parametrics could become more user friendly, more design support oriented with wider applicability.

5 Conclusion

After the review of present day shape parametrics from the view points of underlying technology, major future issues are summarized and some idea for resolving some of issues are presented.
In order to make parametrics a real design tool, further research works and elaborate implementations are required.

6 References

[1] N.Christensen:ISO/TC184/SC4/WG12/N022:Shape Parametrics Framework Proposal,Jul.'97
[2] M.Pratt:ISO/TC184/SC4/WG12/N106: Background Study for Procedural Modeller Interface, Oct. '97
[3] A.Ohtaka:ISO/TC184/SC4/WG12/N109:Commercial CAD example for procedural modeller Interface, Oct. '97
[4] A.Ohtaka:ISO/TC184/SC4/WG12?N189:Parametric Representation and Exchange: Prepararory Knowledge about History Based Parametric Model, May'98
[5] A.Ohtaka: ISO/TC184/SC4/WG12/N295: A Sample Data Model for History-based Parametrics and Key Issues, Jan.'99
[6] V.Shapiro:'Boundary Representation Variance in Parametric Solid Modeling', Jan. '97
[7] C.M.Hoffmann:' On Editability of Feature-based Design', Computer-Aided Design, Vol.27, No.12'95
[8] C.M.Hoffmann:' Generic Naming in Generative, Constraint-based Design', Computer-Aided Design, Vol.28, No.1 '95
[9] J.Kripac:'A Mechanism for Persistently Naming Topological Entities in History-based Parametric Solid Models, Solid Modeling'95,Salt Lake City, Utah USA(1995)

An Overview of the CADCEUS System

Akihiko Ohtaka

Nihon Unisys, Ltd., Tokyo Japan,
Akihiko.Ohtaka@unisys.co.jp

1 Introduction

CADCEUS is an integrated CAD/CAM/CAE/CG system designed to totally support product development processes from conceptual design through manufacturing preparation for automobile, machinery, precision, electric and electronics industries. The term CADCEUS(Computer Aided Design, manufacturing and engineering for Concurrent Engineering by nihon Unisys System) is derived from the English word 'CADUCEUS' which means a stick with two entwined snakes and two wings, which God's messenger Hermes had. It is so named in the hope that it shall strongly support engineers who are engaged in mechanical design and manufacturing.

Manufacturing industries are strongly asked further improvement of product quality, reduction of development cost, and shortening of development time in order to meet wide variety of market needs. Drastic innovations in design and manufacturing are necessary for satisfying these requirements. Key methodologies they are required to tackle are establishment of CIM environment in pursuit of efficient enterprise activities administration by unified information management, and establishment of collaborative and concurrent environment in pursuit of quality improvement and efficient design and manufacturing. CADCEUS is the next-generation integrated system which Nihon• UNISYS has developed to meet above mentioned technical trends and requirements.

• Chapter 2 of this paper describes the objectives of CADCEUS. Chapter 3 and 4 discuss its major characteristics and system configuration, respectively. Chapter 5 outlines major capabilities of CADCEUS.

2 Objective of CADCEUS

Manufacturing industries have introduced CAD/CAM systems for the purpose of shortening lead time of product development. Advantages of conventional CAD/CAM systems, however, have been limited to the• automation and rationalization of individual process. Substantial lead time reduction can not be expected under present circumstances. What are required are not only to improve CAD/CAM systems applied for individual process, but also to effectively combine islands of• automation which resulted from the application of CAD/CAM systems to each process independently. Namely, it is necessary to achieve system integration and generation of necessary enough information transferred or shared among processes. If the information transferred from preceding process is insufficient, it becomes inevitable to reconsider and re-input the information in the current process that has already been defined in the preceding process, which causes double efforts. If this situation is left unchanged, we can never expect productivity improvement, reduction of number of drawings, value shift from drawing to 3D model or realization of concurrent engineering by front loading each process. In the days before CAD/CAM systems were introduced, the intents of preceding process were transferred sufficiently by drawings, instruction manuals, or human-to-human communications. The conventional CAD/CAM systems can mainly transfer shape related information. They cannot transfer design intents under transferred shape, functional features, and various product attributes.

• As far as shape representation is concerned, evolution from wireframe• representation to surface representation, and then solid representation shows considerable progress for representing real object. But, these representations are limited to mathematical description of shape after design. Namely,

design intents of shape are not modeled. This is the reason why drawings are necessitated as before. In order to resolve this situation, "Product Model" which incorporates every technical information of a product during its life cycle has been proposed and discussed among research fields and ISO TC184/SC4.
• The primary objective of CADCEUS is to meet urgent needs of system• integration, unified data management and advanced automation by implementing this "Product Model". It also aims at innovation of the modeling environment, inheritance of design intents, promotion of automation, and incorporation of flexibility to design changes by combining new technologies such as parametrics and form feature.

3   Characteristics of CADCEUS

CADCEUS is an integrated CAD/CAM/CAE/CG system based on integrated database and integrated human interface. It supports all the product development processes from conceptual design to manufacturing preparation and supports realization of Computer Integrated Manufacturing (CIM) and concurrent engineering environments.This chapter briefly introduces product model concept implemented in CADCEUS.

3.1   Product Model

CADCEUS implements product model from the view points of "information integration" and "information completeness."

3.1.1   Information Integration

CADCEUS product model realizes integrated data model of information necessary for all the product development processes for eliminating information loss between processes. It implies that direct use of designed part model is enabled for the study of manufacturability. In conventional environments, there has been cases that different systems are used for successive development processes. In that situation, data transformation between processes is inevitable, which causes loss or deterioration of data. Use of CADCEUS unnecessitate data transformation between processes

3.1.2   Information Completeness

Different from conventional systems, various technical information around product shape representing design intents may be included in CADCEUS product model, which make it easier to understand functional constituents of a product shape and their conditions. Use of these information improves common understanding of the product shape between designers and between processes, and makes it easier to change the product shape according to design changes. The product model with these characteristics in the center, design cycle and manufacturing cycle are performed as shown in Figure1.Earlier manufacturability study by the use of each phase design model enables earlier incorporation of manufacturability requirements, which could lead to drastic reduction of product development time.

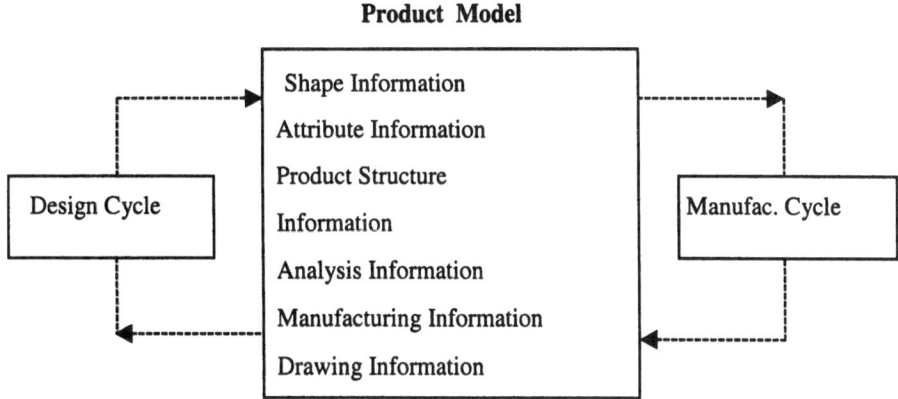

Figure 1: The Situation of the CADCEUS Product Model

3.2 Attributes of the Product Model

CADCEUS product model handles the following types of information:

(1) Shape information
Shape information consists of geometric information, topological information, constructive information, and geometric constraint information. The constructive information and geometric constraint information are used for the capture of design intents. Most conventional systems treat only final information about shape. Therefore, it is difficult to efficiently change shape satisfying design intent. Furthermore, final part shape is represented by a simple collection of geometry segments, which makes it impossible to understand or extract functional units.
In order to overcome these deficiencies, CADCEUS stores constructive information which is the most useful information in CADCEUS modeling. The constructive information contains modeling operation history and modeling condition and reason information associated to shape. These information in conjunction with geometric constraint information realize design intents representation and transfer. CADCEUS shape model allows hybrid representation of wireframe, surface, and solid, which enables set operation between surfaces and solids. CADCEUS parametric modeling is available for this hybrid model, which drastically increases practical usability of parametric modeling. Non manifold representation is also allowed.

(2) Property information
The model can contain material, wall thickness, surface roughness as shape attributes. Parts attributes and machining attributes can be attached to faces for automatic creation of bill of material and for that of machining process planning.

(3) Assembly information
The model can contain information about hierarchical, positional, and connectivity relationships between parts.

(4) Drawing information
Drawing information is related to geometric information so as to realize automatic association between 3D model and its drawings.

(5) Analysis information
FEM or kinematics analysis models, analysis conditions, analysis results can be stored in the model.

(6) Machining information
The model can contain process planning information (machining method, machining conditions, etc.),
cutter location information, and information about residuals.

3.3 Data Structure

The database of the CADCEUS system consists of two concepts for the categorization of design space
and design object. They are "workspace" and "object".

3.3.1 Workspace

The database of CADCEUS is segmented into blocks of appropriate design units. Each of these blocks
is called a workspace. There are two types of workspaces, one is shared workspace(SWS), the other is
personal workspace(PWS). SWS is a common space for two or more designers and PWS is a personal
workspace independent for each designer. A designer may use two or more PWSs. Collaborative
design support is realized by the effective use of these two types of workspaces and data transfer
management between SWS and PWSs.
The concept which corresponds to a product, or an assembly, or a subassembly, or a part, or a sheet, or
a view of some sheet is "object". A workspace may contain arbitrary number of objects.
An object should be closed in one workspace.There are two ways of representing assembly. One is
representing an assembly in a workspace relating objects contained in the workspace. The other is
representing an assembly relating objects in two or more workspaces.
The former is expected to be used in rather earlier design phases and the latter in detail design phase.

3.4 Object

As described above, "object" is a lower level notion of the workspace. Almost all technical data are
contained in "object". Objects are classified into the following four types according to their properties
and uses:
(1) Part: corresponds to a single part.
(2) Assembly: corresponds to an assembly or a subassembly.
(3) Sheet: corresponds to a drawing sheet.
(4) Projection: corresponds to a view(ex. front view) on a drawing.
One characteristics of the object is that it can be instantiated. Figure-2 shows that object-B is
instantiated twice on the object-A. Since all the technical data such as shape are stored in the original
object-B, any change of B could be immediately influenced on its instances on A. The other
characteristics is that objects in a workspace could be geometrically related using hierarchical
relationship or connectivity relationship. Use of the hierarchical relationship enables assembly
hierarchical representation as shown in Figure3

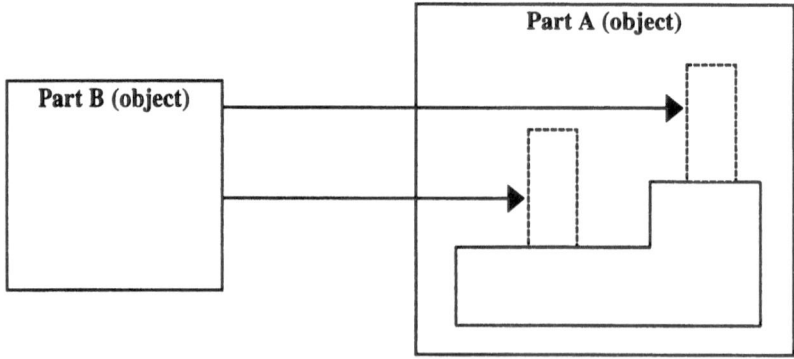

Figure-2: Instantiation of Objects

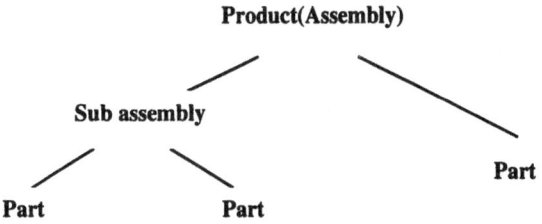

Figure-3 :Hierarchical Structure

Figure-4 shows mutual relationship of different object types, namely sheet, projection, and part objects. All of these objects are related by placement function. If the shape of a part is changed, the change is reflected on the drawings immediately and dimensions on the drawing are automatically changed accordingly.

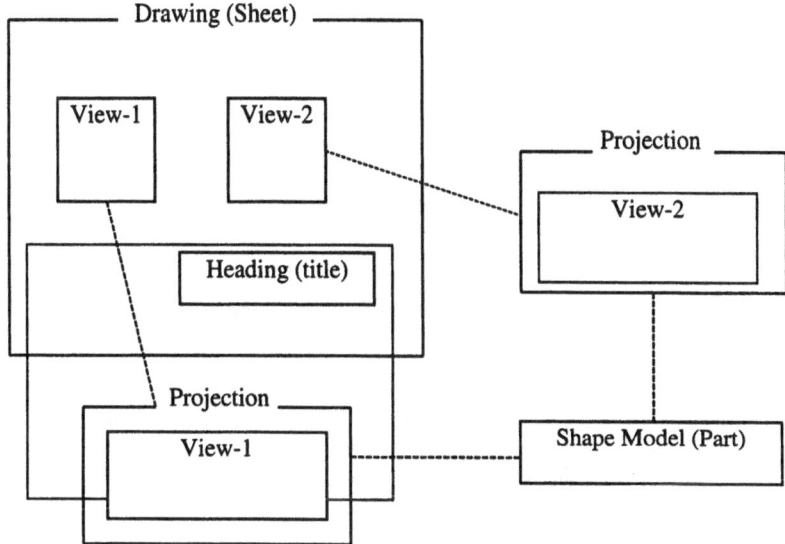

Figure-4: Relationship between Drawing and 3D Model

There are mainly two grouping capabilities. One capability called "group" is used for grouping elements in an object. "group" is used for easy visual control or geometry manipulation. Though the group does not allow hierarchy, an element may belong to two or more groups.

The other grouping capability is "layer". Different from the group, a layer may range to two or more objects. Major usage of the layer is a support of multiple viewing of elements and visual control.

3.5 Collaborative Design Environment

In order to realize reduction of product development time, support of collaborative design work is quite important. CADCEUS supports collaborative design of a product by number of designers by switching private environment and shared environment ,and by maintaining integrity of data in shared environment. Figure5 shows the schematic diagram of the collaborative environment. The two environments are called private workspace and shared workspace respectively. Each designer performs his design in his private workspace without any intervention. He may check in, reference, update, save data to and from shared workspace. All the logical integrity on the latest status of data (object),relationship between objects are maintained in the shared workspace, thus realizing independence of each designer's work and integrity management of all related data among designers.

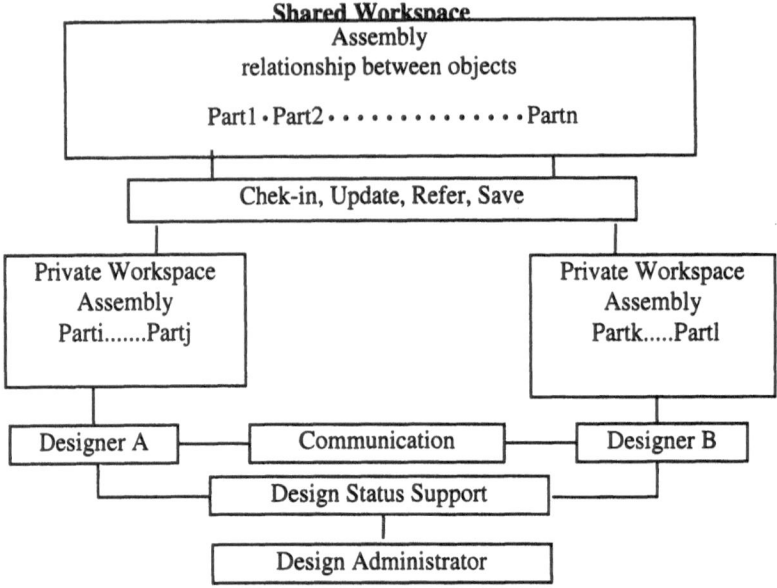

Figure-5: Collaborative Design Support

4 System Configuration

CADCEUS is a typical server/client system that runs on graphic workstations and on personal computers. It realizes centralized data management through a network of host computer, servers, and workstations. The basic centralized/distributed hardware configuration for CADCEUS is as shown in Figure-6.

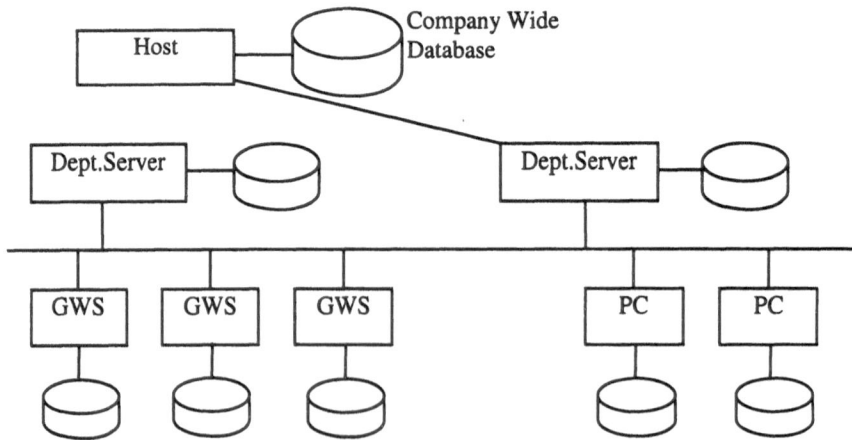

Figure-6: Hardware Configuration

As for the sofware configuration, CADCEUS consists of variet of Subsystems with the following characteristics where the product modeler plays a central rola (see Figure 7).

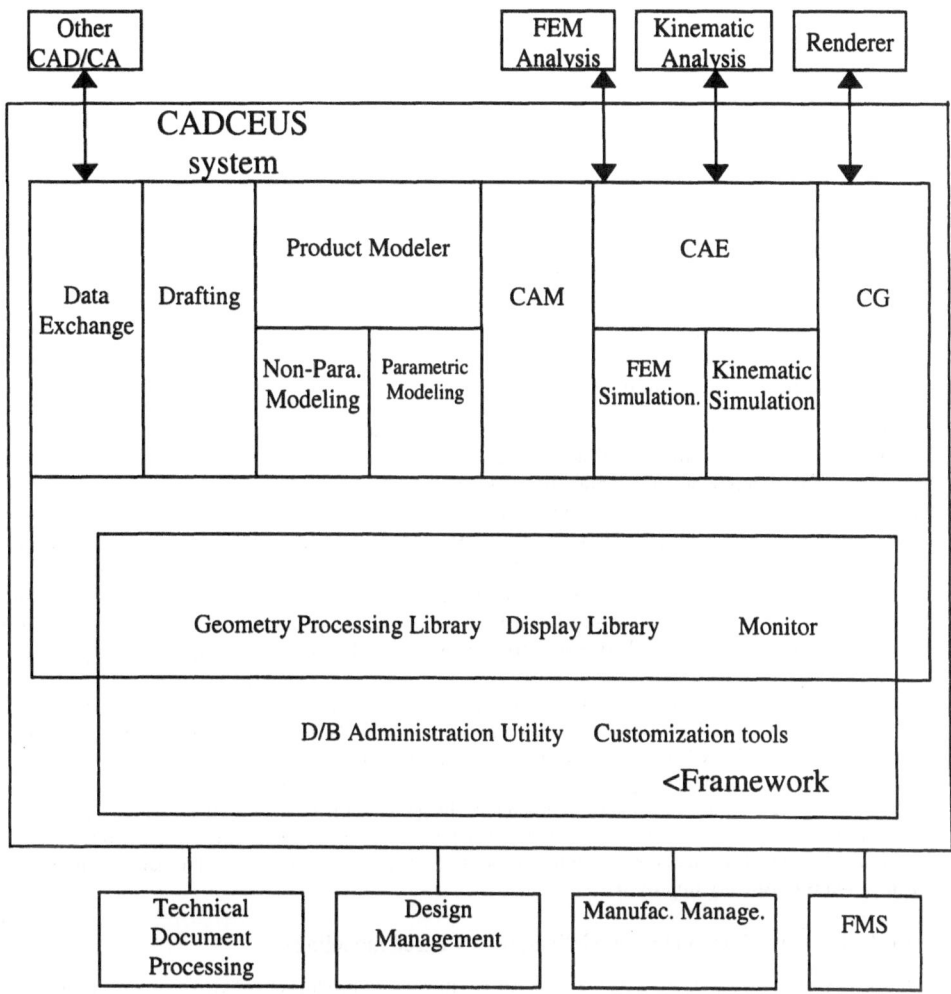

Figure 7: Software configuration

1. Drafting subsystem which has full association between a shape model and its drawings, and has parametric drafting capability.
2. CAM subsystem which supports variety of machining methods and machining planning.
3. CAE subsystem which has variety of pre-processing and post-processing capabilities for FEM analysis and kinematic analysis, and has own solver developed for solving non-linear flow and soridification analysis.
4. CG subsystem which has various capabilities for presentation.
5. Data Exchange subsystem for exchanging data with other CAD systems

In addition to these subsystems which directly support design and manufacturing activities, the followings are also provided.

1)Various utility functions for easing data administration and maintenance

2)Various customization tools for enabling user or activity oriented systems

3)Interface with PDM systems for the integrated management of product data and related documents such as drawings and various technical documents.

5 Functionality

CADCEUS provides various application functions to support product-model based product development processes. This chapter focuses on typical application functions of CADCEUS.

5.1 Shape Modeling

Requirements of shape modeling functions varies depending on the nature of design activities in each process from early design to manufacturing preparation. Examples are design activity where determination of principal sections are crucial, design activity where object shape is designed by combining standard geometry, layout design of parts, detailed modeling of die faces, die structure modeling and so on. It is not realistic to satisfy variety of requirements by a single modeling approach. CADCEUS supports combination of the following modeling functions.

5.1.1   Geometric constraint modeling

Different from the conventional method of directly representing target shape as a collection of geometric elements, geometric constraint modeling is a method in which the system memorizes geometric constraints which the target shape must satisfy and variable parameters such as length and angle, and various relationships among parameters. When values are given to parameters, the system automatically generates shape satisfying all the prescribed constraints and the relationships among parameters. Constraint conditions include horizontal, vertical, parallel, perpendicular, length, and angle. CADCEUS is provided with a sketching function that facilitates the definition of geometrically constrained model. As an extension to geometrically constrained part model representation described above, parametric positioning between parts which is usually called'parametric assembly' is also supported.

Compared with conventional modeling methods, geometric constrain modeling has strong advantage of being able to store design intents directly into the model. This capability is especially useful in those design activities where number of sectional geometry or fundamental shape should be studied with variety of conditions, design of slightly modified part shape, design activities where registration and repeated use of standard shapes are dominant. Shape model generated by this method can be used as a profile of solid or surface primitive, or constituents of drawing. In combination with the procedural parametric modeling capability described in the next paragraph, automatic shape model generation is drastically improved.

5.1.2   Procedural parametric modeling(History based modeling)

In every product development processes, needs for modifying product shape frequently occur in order to meet the study of alternative design plans or to meet change of design. Procedural parametric modeling method strongly support these environments. In this method, the system memorizes shape modeling procedures and conditions. When change of parameters, for example change of dimensions, or change of procedure itself occurs, the system automatically regenerates necessary portion of the given model by re-executing memorized procedures with given conditions. This modeling method can be applied not only for solid modeling, but for surface modeling, wireframe modeling or their hybrids.

5.1.3   Form feature modeling

Form feature is a group of geometry which correspond to a functional unit that designers recognize. Typical features include protrusion, rib, fillet, and hole. Form feature modeling allows modeling operations not with conventional geometric segment level but with form feature level.

In• conjunction with procedural parametric modeling described above, shape modeling preserving form feature is made possible. Form features are application dependent by nature. It means that predefined form features are not sufficient for various applications. The user of CADCEUS can easily define own form feature by registering operation history for creating it.

5.1.4   Hybrids of conventional modeling and advanced modeling

It is understood that not all product shape is appropriately or efficiently modeled with the mixture of advanced modeling methods such as geometric constraint modeling, procedural parametric modeling and form feature• modeling. Based on this understanding, CADCEUS also supports conventional modeling with sophisticated capabilities such as automatic trimming of surfaces, set operation of surfaces with solids, automatic surface model generation based on wireframe model, automatic fillet generation, compound offset surfaces, spring back estimation, overcrown estimation, wide variety of drafting angle estimation capabilities etc.

Compound surface offset is specifically powerful. It can be applied for not only shelling but also for variable offset of given surface model. Typical application of the variable offset is automatic creation of rear faces of a bumper. After the completion of front surface model of a bumper, designer only specifies offset values at necessary points, then the system automatically completes rear surface model by offsetting constituent surfaces, recognizing topological inconsistency after offset such as intersection, overlap, gap, disappearance of edges or even faces, and automatically resolving those topological inconsistencies.

5.2   Die packages

CADCEUS is provied with domain specific application packages in automobile die design / manufacturing area.. In design area, CADCEUS provides dieface design package, stamping die structure design package and mold die structure design package. There are three focuses in these packages. The first is use of hybrid modeling and parametric modeling for re-usability of data and efficient design. The second is incorporation of design criteria and know how into the system, and the third is attachment of various attributes to faces so that later automatic creation of bill of material or aotomatic creation of machining process planning are made possible.

In manufacturing preparation area, CADCEUS provides DieCAM package which is a solid/ parametric based die structure CAM. Though DieCAM can be used independently, combined use of it with die design packages realizes pure integration of CAD/CAM in die area with maximum effect.

5.3   Drafting

The features of the CADCEUS drafting capabilities are as follows.

5.3.1   Flexibility on various drafting standards

CADCEUS supports drafting in accordance with various drafting standards such as JIS, ISO, and ANSI.

5.3.2   Flexibility of units of measurement

CADCEUS allows dimension presentation in various units of measurement in accordance with those used in shape representation.

5.3.3   Associationy between shape model and drawings

In the case of 3D shape model based drafting, CADCEUS handles each view• of a sheet as a projection of the 3D shape model and memorizes the relationships between the 3D shape model and each projection. Manipulation of these relationships enables automatic reflection of shape model change on drawings. CADCEUS also allows independent visibility control of geometric elements in each view. Examples are change of line styles and visibility• on/off control. It supports an automatic hidden line elimination function for solid models.

5.3.4   Parts list generation

CADCEUS allows the user to create a parts list by making use of table manipulation functions. The number of parts could be automatically calculated from the hierarchical structure of the model.

5.3.5   Parametric drafting

CADCEUS allows the user to perform parametric drafting by combining geometric constraint modeling and conventional wireframe modeling functions. It means that dimension-driven parametric drafting where changes of the shape model on the drawing sheet satisfying all the prescribed constraints are automatically performed in accordance of the dimension changes

5.4   Computer Aided Manufacturing (CAM)

CADCEUS CAM subsystem supports various machining methods such as machining parts of complicated free form surfaces, drilling, and area machining of raw material, etc. The other features of this subsystem are as follows.

5.4.1   Machining planning

Capabilities for easing users definition of machining portion, machining range, machining methods and machining conditions are provided.

5.4.2   Cutter location calculation

This subsystem enables efficient machining for profile milling and character milling by automatically calculating cutter locations taking into account residuals of the previous machining process throughout the processes from rough milling to final finishing.

5.4.3   Treatment of machining know how

This subsystem allows the user to store various machining related technical information such as machining procedures and conditions. By the use of pre-registered information, the user can efficiently plan current machining.

5.4.4   Machining Simulation

This subsystem is provided with computer simulation of real life machining. By the use of this capability, machining quality such as resudual or overcut is precisely investigated before the real life machining and optimal machining planning is made possible.

5.5   Computer Aided Engineering (CAE)

One of the features of the CAE pre/processor is an effective use of form feature and parametric functions for easily obtaining idealized shape for FEM analysis. Namely, very small holes or fillets which do no affect analysis results can be temporarily deleted by the use of above functions.

As for analysis model generation, automatic mesh generation capabilities for surface model and solid model, and adaptive mesh capability are provided.

As for analysis programs(solver), the user can select the most appropriate program for his application from among Unisys proprietary analysis programs MELTFLOW(resin flow analysis), CAST(resin solidification analysis), and METALFILL(metal flow analysis) or commercial analysis programs such as NASTRAN, MARC, ARGUS(nonlinear analysis of sheet metal), Applied Structure(sensitivity analysis, optimization analysis), and JOH-DYNA(shock analysis).

As for the post-processor, variety of animation functions in addition to the conventional output forms such as deformation diagrams, contour diagrams, arrow diagrams, and XY-graphs are provided.

5.6 Computer Graphics (CG)

CG subsystem mainly supports conceptual design by advanced rendering capabilities such as ray tracing or texture mapping ,by painting functions such as image composition, and by various multi-media functions.

5.7 Data Exchange

Data exchange subsystem of CADCEUS consists of processors that provide interfaces listed below. They are organized around the intermediate file as shown in Figure-8.

(1) STEP interface

(2) IGES interface

(3) JAMA-IS interface(JAMA: Japan Automobile Manufacturers Association)
(4) Direct interfaces to major commercial CAD systems(CATIA, Pro/E, UG, I-DEAS)

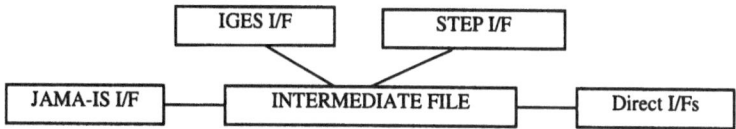

Figure-8: Data Exchange Subsystem

5.8 Customization

CADCEUS is provided with customization capabilities which allow the user to establish his own operation environment, to add his own functions, and to promote automation. Major capabilities are;
(1) Modify screen layout
(2) Alter and reconstruct menu layout
(3) Change messages
(4) Add users own commands
(5) Add users own subsystems

5.9 Inter-operability with third party systems

In order to obtain inter-operability with third party systems, CADCEUS is provided with a mechanism called Open APIs. This allows third party systems freely look into the latest status of CADCEUS data base. Typical examples of the usage of Open APIs are, combination of CADCEUS with DYNAMO/Tecnomatics, VisFly/EAI.

6 Conclusion

This paper briefly summarizes major characteristics and capabilities of CADCEUS which are expected to contribute to drastic improvement of design environments ,and promotion of automation. The practical utilization of colaborative design support environment and advanced modeling capabilities (such as parametric, geometric constraint, form feature) has just begun recently. Practical experiences of these capabilities will clarify real needs in actual design and manufacturing environments. We intend to further sophisticate these capabilities based on those needs.

Digital Processes in Product Creation

Gustav J. Olling

DaimlerChrysler Corporation, Auburn Hills, Michigan, USA
gjo@daimlerchrysler.com

Abstract: Today, as many emerging technologies are enhancing the efficiency of the product development and manufacturing processes, a major concern is the strategy to be used in automating and integrating a manufacturer's operations with these technologies. This presentation shows how the adoption of a core CAD/CAE/CAM/IS strategy at DaimlerChrysler AG has facilitated the integration process. It also discusses the technological developments that are occurring throughout the company. It closes with a sketch of the evolutionary stages of corporate integration and a summary of conclusions drawn from the integration process.

As the 20th century comes to an end, political and economic forces are moving nations towards the formation of a global community. The easing of tensions between major powers, the forging of broader alliances, the growth of trade to worldwide markets, expanding travel and electronic commerce, the streamlining of businesses, pervasive automation, and the adoption of flexible methods have transformed the practice of manufacturing and the direction of research and education in manufacturing. In pursuit of world-class manufacturing status, global firms have realized that a commitment to quality and responsiveness to customers' demands are now the benchmarks of outstanding enterprises. Industry leaders surpass their competitors by launching rapid product introductions that offer the highest attainable quality at the lowest available price. World-class manufacturing has become a process of continuous improvement, focusing on the need to reduce lead-time while continuing to innovate and assure quality.

Innovation, or the conversion of great research into great products, is the first rule of international industrial competition these days as companies must respond to shifts in global markets, quickly and effectively, by delivering new products in the shortest time, with the highest quality and the lowest cost. The successful enterprise of the next century will be characterized by an organizational structure that deftly responds to customers' demands and changing global conditions, no matter how large the enterprise becomes. It will create a corporate culture that empowers employees at all levels and facilitates constant communication among related groups. It will have a technological infrastructure that fully supports process improvement and integration despite unexpected changes in corporate direction. To keep up with rapid developments in global manufacturing, the enterprise must first look at its organization and culture, and then at its supporting technologies. Too often, companies throw vast capital at technology to catch up to competitors only to find that their employees are still clinging to the same attitudes, methods, and timetables that caused problems in the first place.

That being said, technology is certainly of prime importance in manufacturing today because of its potential for having an immediate impact on key factors such as productivity, quality, and cost. Designers and engineers now use computer-aided design and engineering (CAD/CAE) tools to create three-dimensional images that can be textured and reflected, analyzed, and optimized to refine the component design and then assembled and disassembled in digital mock-ups (DMU) to evaluate manufacturability and serviceability. Manufacturing engineers use computer-aided manufacturing (CAM) systems for process planning, tool design, machine programming, and virtual assembly lines. Administrators draw upon corporate information systems (IS) in their production planning, production and inventory control, and sales research and forecasting.

Virtual product and process modeling is the ultimate goal in integrating CAD, CAE, CAM, and IS technologies. All of these key areas have to be addressed to completely systematize the manufacturing process. Each area has a distinct set of product types, technical trends, and potential benefits. It must be realized, however, that the expected gains from these areas depend upon an intelligent balance of all the technologies and that maximum benefits come from the integration of the technologies into the overall manufacturing system, which comprises activities from design intent through manufacturing planning to actual manufacturing, including quality control, shipping, and customer feedback. The integration involves the vertical and horizontal integration of all these activities. With such integration, all of the major areas of manufacture can work at a common pace towards a common end. Without such integration, the less advanced areas will continue to hamper, and in some cases nullify, the efficiencies of the more advanced areas.

The strategy of developing product design and production processes side-by-side in a digital environment is known as "concurrent" or "simultaneous" engineering. The goal of this strategy is to completely integrate the engineering, manufacturing, and management functions throughout the product development process, taking into account the manufacturability, assembly, and serviceability of the product. The technologies required for this kind of engineering create complete product models and process-definition models that integrate business, part, and process information, manufacturing and assembly process design, plant design, and product serviceability requirements. Needless to say, the achievement of such complex models, thorough automation, and constant tracking and communication of information requires a great deal of commitment, capital, training, and time. Nothing about automation is automatic.

Crisp, accurate, real-time communication (with "technology flow-through") throughout the extended enterprise is the cornerstone of the virtual product creation process. Technology flow-through – for example, digital mock-up technology – is being used not only for virtually assembling an entire product, such as a vehicle, but also for designing and simulating an entire machining line, including resources such as stations, fixtures, tools, material handling, and ergonomics. The focus of activity in the integrated manufacturing environment is the product model. To carry information related to the diverse processes of manufacturing, the product model must have a large capacity for information but a small enough size to make it portable throughout the product life cycle. An expansive scheme of data representation, such as STEP, allows for the integration of diverse technological and organizational approaches without forcing uniformity on the enterprise's computer systems. However, such schemes are not yet available in production-ready forms. The best current solution, therefore, is to implement a core-system plan in which all of the major areas of the company use the same set of integrated CAD/CAE/CAM/IS tools, communicate over a single network, share a common database, and adhere to standard design practices.

The chief advantages of the core-system scheme are that it prevents the loss of design intent through the imperfect translation of data between systems, eliminates the preparation of data before translation and the cleanup afterwards, and facilitates the implementation of concurrent engineering in product development and manufacturing divisions. At DaimlerChrysler AG, for example, we are promoting a core-system solution to tighten the integration between diverse areas of the company and between the company and its international suppliers. A major concern is to minimize translation. Since competing software systems usually have different approaches to data representation, translations between systems make arbitrary interpretations and compromises. In sensitive applications, such as the design of the outer surfaces of sheet metal, translations do not preserve the original intent and integrity well enough to produce a satisfactory result. Eventually, the company would like to eliminate all intermediate forms – not only models from other CAD systems and the IGES files used to translate them, but physical bucks and mock-ups, clay models, and background CAD models. Experience has shown us that intermediate forms add delay and cost, cause synchronization problems, and lead to quality shortcomings.

A video was presented at the conference to illustrate the application of digital CAD/CAE/CAM technology throughout the vehicle development process at DaimlerChrysler. The process starts with conceptual design, proceeds through body engineering, digital model assembly, powertrain design, and rapid prototyping, and finishes with stamping and assembly. In the early stages, a conceptual sketcher is used to create a 2D sketch of the design concept. This sketch is converted into a rough 3D computer model that can be rotated in real time for preliminary visual inspection. After the design has been refined into a more finished 3D CATIA® model, it can be evaluated for aesthetics in a number of ways. Realistic rendering systems can create a high-definition animation of an entire vehicle overnight and display the full-sized result for a senior management review by the next morning. Realistic rendering enables designers to evaluate more alternatives for exterior features such as fascias, wheels, and grills in a limited time. Another application, virtual reality (VR), is used to analyze vehicle interior designs. Designers change interior surfaces in CATIA® and then generate a new VR model. Evaluators can then navigate through the design to judge whether it has the desired look and feel. External surfaces are also evaluated with reflection lines. Using CATIA® macros developed by DaimlerChrysler, surface designers can cast lines of light on curved surfaces to assure that the curvature is pleasing and free from anomalies. In inner and outer sheet metal panels, engineers can go a step further to evaluate the parts for manufacturability. Deformation software analyzes how sheet metal will behave under the stresses of stamping on the factory floor and slamming in the customer's garage. Ensuring that the metal will deform as expected enables the manufacturer to produce better fit and finish in external surfaces and to improve noise, vibration, and harshness (NVH) characteristics.

In the engine compartment, the entire engine assembly is designed in CATIA®. Once the components of the engine are designed, they are analyzed with CATIA® and third-party software. For example, the intake manifold of the 2.7L engine was subjected to fluid dynamic analysis to verify the air flow characteristics of the design. After 1,500 design iterations, the analysis and redesign produced an engine with 10% greater fuel efficiency and 30% fewer emissions. These iterations took six months less time than traditional methods and saved up to $100 per vehicle on catalytic converter materials. Another focus of fluid dynamics study is the flow of the air/fuel mixture through the combustion chamber. Positioning the flow with respect to the spark plug helps to improve performance and reduce emissions. The study of the air/fuel mixture can be followed all the way through the vehicle, starting at the intake ducts, continuing through the intake into the cylinder during combustion, out the exhaust manifold, through the catalytic converter, and out into the world. At each stage, the flow of gases can be optimized to gain the highest efficiency.

In the virtual factory, we can simulate many of the manufacturing processes to test feasibility before any physical setup is performed. Besides testing sheet metal designs to prevent tearing during the stamping process, we can create 3D models of the tools that make the products to see how they fit and interact with the products during manufacture. We can also use generative numerical control technology to mill out foam patterns of dies. This automated creation of the die design from the part design reduces development time from five days by traditional methods to just five hours. In addition, it dramatically improves the accuracy of the physical dies. In the area of process modeling, DaimlerChrysler is presently incorporating the Digital Manufacturing Process System (DMAPS) at its facilities. This system defines and simulates the build sequence as it will take place in the plant, and it does so up-front in the product development process while the product itself is being designed. A further extension of the product design to process design lies in the automation of plant construction. At the Bramalea plant in eastern Canada, the virtual construction was performed by outside architects in a pilot study done in parallel with the 2D design of the plant. The project simulated seven tooling stations, a safety fence and walkover, manual load platforms, electrical equipment, part racks and bins, and electrical raceways. Although the results were not perfect in practice, the studies were continued at Bramalea and then extended to a new, larger facility in Toledo, Ohio. This more ambitious project, involving eight outside vendors, encompassed a facility of over a million square
feet – which is to say that the company had a great deal of faith in computer-assisted plant design. One of the most interesting aspects of the project lay in interference checking between ductwork, piping, structures, and process equipment, which found 202 interferences between these structures leading to about $1.3 million in material cost avoidance.

The core system of DaimlerChrysler's vehicle development process, CATIA®, by Dassault Systemes, is shown in an illustration of the "CATIA® Pipeline" (**Figure 1**). The CATIA® Pipeline was adopted as a corporate metaphor to show the unity of technology flowing through the core system, but also to show the diversity of technology that can be incorporated into or integrated with the core system. The aesthetics of conceptual design and body-in-white styling must be accommodated and, likewise, the intricacies of digital model assembly, powertrain engineering, and wiring schema. The broad

118

G. J. Olling

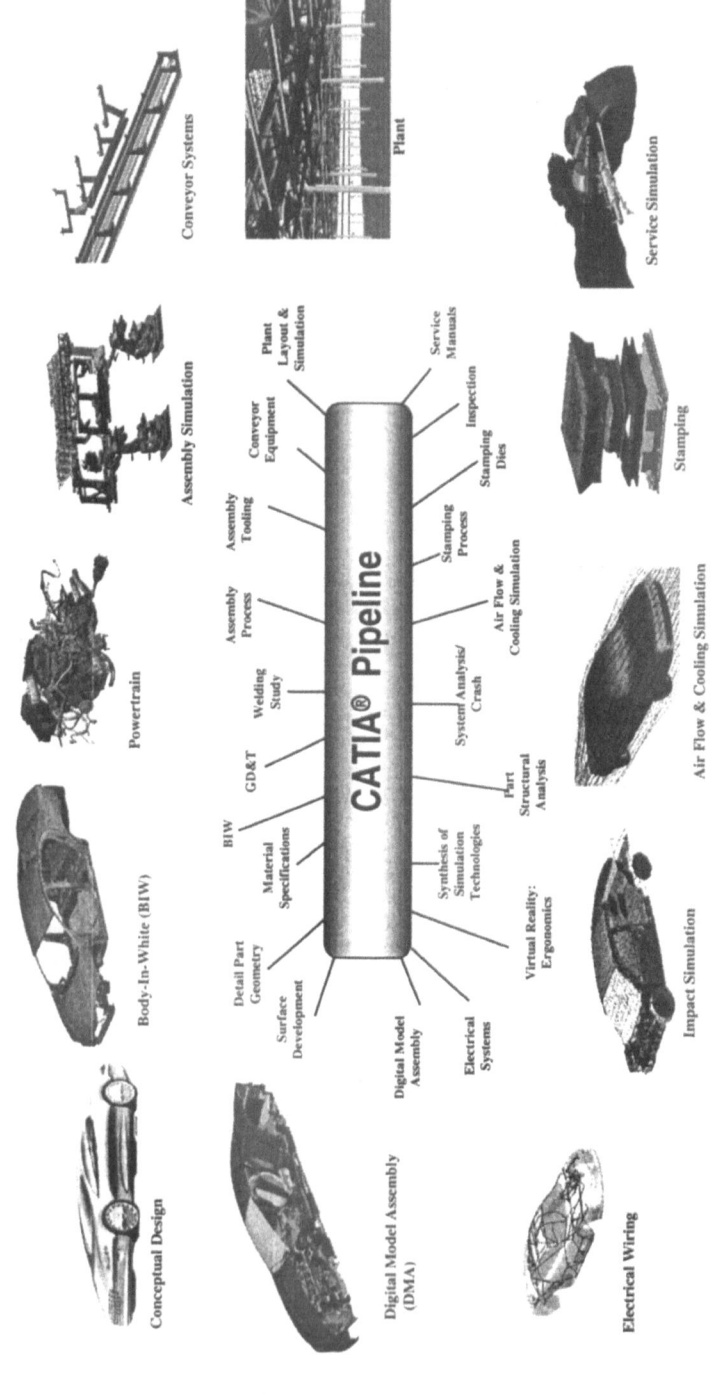

CATIA® is a registered trademark of Dassault Systemes

Figure 1.

strategies for plant layout, stamping dies, and conveyor systems must be supported, but, too, the mathematical dynamics of simulating impact, fluids, assembly, and repair. Such diversity calls for a mature and robust core system that expands with the needs of its major customers.

The current developments in the CATIA® Pipeline at DaimlerChrysler are many. In conceptual design, we are working on sketching and the 2D/3D paint box process, generating 3D surfaces from 2D sketches, photo-realistic rendering in real time (not overnight), and numerical control machining for validation by physical property development. In body-in-white, we are beginning to use a generative shape modeler (GSM) and feature-based design technology that will further automate the activities of part creation and modification. Powertrain engineers have implemented a solid modeling design process on several new engine types. Electrical wiring specialists are using software to develop electrical systems requirement specification, integration between schematic and E3D wire routing, and automatic pathway definition. Digital model assembly groups are examining assembly management techniques, simultaneous engineering, visual analysis of fit and finish, ergonomic simulation using virtual reality, and intranet issue tracking. Impact simulation experts are developing a front, side, rear, and barrier offset system, simulations of structures and occupants, and static crush simulations of doors, roofs, and seat belts. In air flow and cooling simulation, we are working on engine compartment thermal analysis, combustion analysis, powertrain air and fluid flows, heat/ventilation/air conditioning (HVAC) performance, and noise management.

Much development is also in progress in manufacturing and service. Stamping engineers are looking into draw die development, digital process planning, generative feature-based die design (VAMOS), generative pattern construction, and generative 2D and 3D pattern machining. In assembly simulation, we are pursuing digital process definition (SCOPES), fixture design with parametrics, digital workcell definition, robot simulation and off-line programming, and assembly process simulation. We are also implementing an integrated conveyor design package in CATIA®, comprising 3D routing and layout, design and detailing, and definition of the installation and maintenance processes. Plant facilities engineers are exploring integrated 3D plant design, construction management, concurrent engineering, and interference checking between structure, facility, tooling, and equipment. Finally, in service and repair, we are formally defining the service process and developing software to generate removal paths, analyze part, tool, and technician clearances, optimize service factors, and automatically generate service illustrations.

A major advantage of using a core system throughout the product development process is the leverage gained in technology flow-through: a technology created for one task or discipline can often be used in other applications. For instance, the same surfacing or morphing technology that we created for conceptual design is used in body sheet metal design and stamping design, as well as other areas. The digital model assembly (DMA) technology that is used to "fly through" an engine is also used to fly through the complete vehicle, the assembly tooling, and a 3D model of the factory. Simulation technology developed for modeling vehicle impacts is also used for studying sheet metal deformation during stamping.

With a core CAD/CAE/CAM/IS system in place, DaimlerChrysler's next challenge is to incorporate business, engineering, manufacturing, and supplier know-how into a single integrated system. A "next-generation" core system will be based on precise solid models with a feature-based data structure that stores information related to all steps in the product development and manufacturing processes. The system will have a parametric associativity so that design changes in one aspect of the product are automatically reflected in all other aspects of the product. Finally, the system will be generative. Engineering analysis models, manufacturing process models, business cost models, and purchasing materials models will all be generated from the base product design model through built-in intelligence. This kind of generation will be possible because the product model will no longer be geometry represented in the computer but rather will be a section of the online product database.

While the DaimlerChrysler concept of integration uses the metaphor of a pipeline, the company's strategy of expanding automation is visualized as a product and process development "machine" (Figure 2).

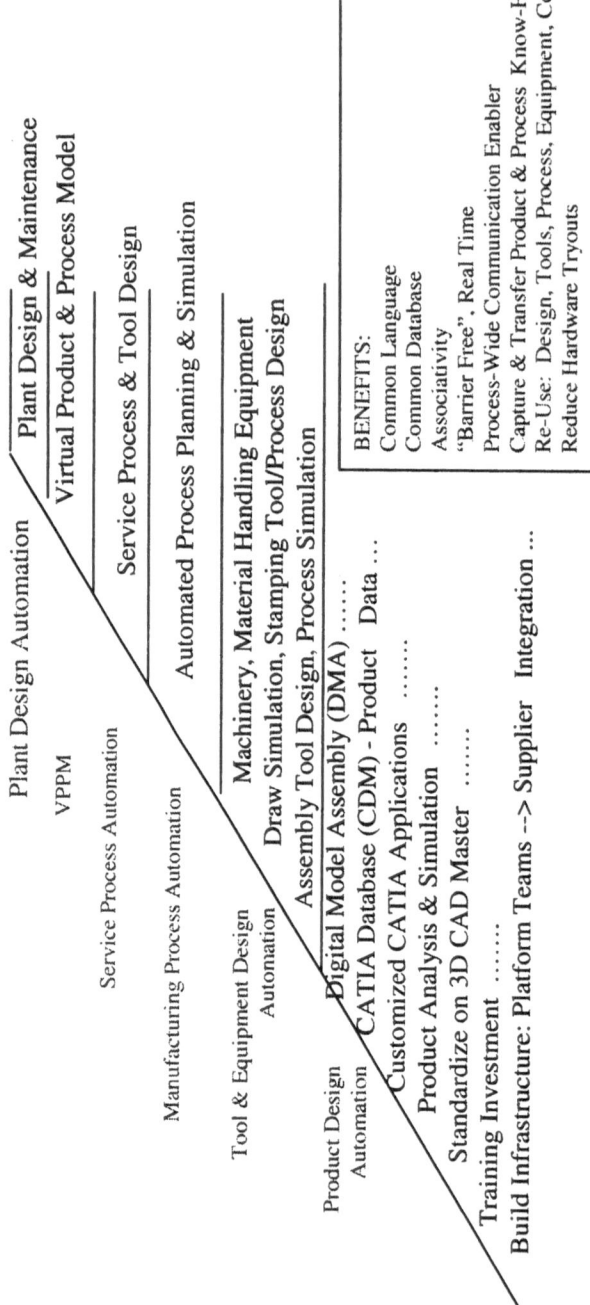

Figure 2.   Automation Strategy

This strategy is evolving through five stages:

1. Product Design Automation

2. Tool and Equipment Design Automation

3. Manufacturing Process Automation

4. Service Process Automation

5. Plant Design Automation

When this plan was conceived about 10 years ago at the former Chrysler Corporation, the goal was simply to automate and integrate the company's internal processes. Since then, however, with the maturation of CAD/CAE/CAM/IS technologies, the increased acceptance of international standards, the opening of global markets, and the company's merger with DaimlerBenz, the goal has been enlarged to read "to automate, integrate, and globalize". That is, effort is invested to adopt broad-based product features and international standards such as STEP; design, engineering, and manufacturing systems that communicate with the outside world as well as with each other; "departments" that address all the aspects of a single product rather than one aspect of many products; and employees who play multiple roles and understand product requirements outside their own areas. Now, as DaimlerChrysler reaches each stage of internal integration, it prepares that stage for global integration.

Manufacturers throughout the world today are striving to automate and integrate their operations to improve the overall efficiency of their operations, even as they expand into global markets. As they grow, they want to add muscle, not fat. Companies have taken many different approaches to modernization. Only one approach, that of DaimlerChrysler, has been considered here. The DaimlerChrysler approach has been to adopt a core CAD/CAE/CAM/IS system and to extend automation and integration from that core. Other successful companies have obtained the best available technology for each area of the company and then have sought out the best integration tools to bind the areas together. While there is no single "best" way to succeed, certain conclusions have become clearly evident by this time. If a set of computer systems is to become the core of an enterprise's business operations, the systems had better be well developed, stable, integrated, and extensible. Second, the organization of the enterprise has to change to take advantage of the more automated and integrated operations. Employees have to be trained in the new technologies, certainly, but they also must be able to focus concerted effort on each product so that they learn from each other, in real time, as they develop and manufacture the product. Finally, the management of the enterprise must not think of the company as an isolated entity. It must reach out to related companies, its design and component suppliers, its software and services vendors, educators and researchers in nearby communities, customer groups, and professional organizations. Only when a company extends its relationships to such a broad expanse can it hope to be truly "integrated".

This paper was first published in "VDI Berichte 1489", ISBN 3-18-091489-0, at the Conference "Virtual Product Development in Vehicle Engineering" by the VDI-Gesellschaft Fahrzeug-und Verkehrstechnik", 09. and 10. September 1999 in Berlin, Germany.

Implementing STEP for the exchange of CAD Data: Some first hand experiences

Markus Schichtel

BMW Group, Munich
Markus.Schichtel@bmw.de

1 Introductory Remarks

This contribution is not intended to give a technical overview on STEP (STandard for the Exchange of Productdata). Rather, the goal is to share some of the key experiences the author has collected during a period of almost four years (1995-99) implementing a STEP processor for the exchange of CAD data between volume modellers and being a representative at the vendor round tables at ProSTEP and PDES, Inc. [1]
Furthermore the major experiences gained during the PICANT project at BMW will also be included.
These key experiences will be organized in four concise statements substantiated by a sound explanation.
Again, a technical exposition of STEP is not intended by this paper. Nevertheless, a few words on some fundamentals on STEP shall be said in the following section to better understand the line of arguments that leads to the statements the reader will be exposed to later on.

2 Some fundamentals on STEP

To the best of the author's knowledge STEP is the first and possibly only standard so far for the exchange of product data, and CAD data in particular, that is firmly based on sound objectoriented data modelling. As a consequence the exchange of data via STEP is based on a data schema. This fact alone represents in the author's opinion a major breakthrough compared to other standards like IGES or VDA-FS and is the main reason for STEP being a powerful tool for data exchange.
The data modelling language chosen, EXPRESS [2], allows not only to model data in an objectoriented fashion, but also to give data semantic meaning by allowing the user to formulate local, socalled WHERE rules that reside within the definition of an object or global rules that specify   the behaviour of relations between objects within a data schema.
Hence, each STEP file can be viewed as an instantiation of a given data schema on which the exchange is based on. Different schemata give rise to different data exchanges via STEP and different requirements for data exchange can be captured by different schemata. Consequently one cannot really regard STEP as just one single standard. It is by far more: It is a technology to create standards! This will become clearer as we move on in this section, but this fact is very important to keep in mind.
The fact that data exchange via STEP rests on objectoriented data modelling principles is directly mirrored in the overall design of most STEP translators. Typically a STEP translator consists of three parts: A scanner/parser to read and write a physical STEP file, a temporary STEP database that buffers the data based on a given schema, and a converter that establishes the link between the STEP database and the CAD system.
In many cases the scanner/parser and the STEP database are provided by third party software vendors to CAD system vendors in which case only the converter lies in the sole responsibility of the CAD vendor. However, in some cases complete STEP translators are offered by third parties or, vice versa, although rarely, the CAD system vendor covers also the development of the scanner/parser and the STEP database.

Now, STEP has been structured into many building blocks called "parts" (which has nothing to do with parts being designed with the help of CAD systems), in order to manage the shear amount and complexity of data involved in the exchange of product data.

The most fundamental ones besides the standardization of EXPRESS and the syntax of a STEP file are the socalled ressource parts (40series) of which the most important one from a CAD system point of view is part 42: Geometry and Topology.

Based on the ressource parts the exchange of data via STEP can be tailored to suit various application domains. This tailoring gives rise to the various socalled Application Protocols (AP) (200series) of which AP203 (Configuration Controlled Design) and AP214 (Automotive Design) are the relevant protocols for the exchange of CAD data today.

Each ressource part and each AP is described by its own unique data schema. Well, this is slightly stated incorrectly as in the case of APs we actually have to deal with two schemata: we need to distinguish between an ARM schema (Application Reference Model) and an AIM schema (Application Interpreted Model). The link between the two is provided by a socalled mapping table, which is an integral part of the definition of an AP.

Typically the user community of an application domain thinks in terms of the ARM whereas the AIM is the foundation of implementing STEP based data exchange. In other words, an implementer is mainly concerned with the AIM since a populated STEP file is based on the AIM.

As a matter of fact, one could easily create an AP for particular needs. For example, one could come up with an AP to describe the exchange of documents between Document Management Systems. All that is needed would be to agree on a particular schema. The scanner/parser and STEP database software is generic enough to handle any STEP file as long as the schema on which the STEP file is based on is known to that software. All you need to do would be to provide the conversion part in order to build a STEP translator for the exchange of documents.

Therein lies the true power of STEP and it shouldn't be difficult by now to follow the conclusion that STEP is not just another standard but rather a technology to create standards.

Now, each ressource part and each AP is subject to its own promotion and approval process starting from a socalled Committee Draft (CD) via a Draft International Standard (DIS) until it reaches the status of being an International Standard (IS). Each stage requires a socalled ballot cycle among all nations being represented in ISO which makes the introduction of STEP slow. To complicate matters a STEP part in status IS may be revised after a time freeze period of five years and reenter the whole promotion process. Since the various parts progress with various speed towards IS the consequences are far reaching and affect tremendously the development and marketing strategy of STEP translators. We will pick up on this later on.

To summarize the reader should get acquainted with the following aspects fundamental to STEP

STEP is firmly based on objectoriented data modelling techniques

STEP is NOT a standard but rather a technology to create standards

STEP parts are subject to formal approval processes through ISO

STEP is not monolithic but a set of standardized building blocks in various stages of approval

3   Key experiences with implementing STEP for CAD data exchange

The first statement is based on the daily work of programming a data exchange processor based on STEP. It can be said that

Statement 1: STEP is not necessarily implementation friendly

but rather a tedious and difficult task which sometimes requires extensive rework and iterations.

To begin with, a huge amount of documentation needs to be digested compared to older standards like VDA-FS or IGES before a software engineer can seriously consider starting to architecture a STEP translator or even write code. To give the reader an idea it shall be stated that VDA-FS required just about 30-40 pages while IGES specifications come as a fairly comprehensive book of about 300-400 pages. Though this is already one order of magnitude more, with STEP we talk about yet another order of magnitude more compared to IGES. The specification of AP214 alone comprises roughly 2800 pages. Each ressource part comes as a document of about 150-200 pages. In addition one has to cover the data modelling language EXPRESS. It is obvious that a STEP novice is definitely in need of STEP experts to get a considerable amount of consulting in order to understand the concepts behind STEP.

Consulting by STEP experts is also necessary to find out how to map certain concepts of a CAD system (eg. a specific type of geometry or structural information like assembling solids) to a STEP file. Consulting is often hindered because STEP modelling experts are used to think in terms of the ARM whereas an implementer is only interested in the necessary AIM entities. In many cases a STEP implementer believing that he has found a solution in terms of the AIM is not aware that he may violate rules on the ARM level or that a mapping from ARM to AIM is lacking for his solution.

Furthermore the STEP documentation is imprecise or leaves room to interpretation. Important information is often hidden in innocent IPs (Informal Propositions) and hence could not be found easily. In light of the fact that two APs are involved, rule conflicts between AP203/214 had to be dealt with as well.

Another reason why implementing STEP can be tedious is the way in which the STEP database is accessed. This is done through the Standard Data Access Interface (SDAI) which is a rather low level interface. SDAI does not provide access at the level of AIM entities, only at the level of its attributes. SDAI gets quite clumsy when it comes to arrays and matrices of numbers needed eg. for the exchange of control points of NURBS. Consequently, wrappers are needed that provide an interface between the converter of a STEP translator and the STEP database. Writing these wrappers is a really time consuming task.

Fortunately for about 80%-90% of all AIM entities needed wrappers can be produced by a code generator. However it takes a while until a STEP novice would become proficient enough to build such a code generator.

It is true though that socalled High Level Interfaces (HLI) are provided commercially but they are bound to one specific schema. All major CAD systems however need to support more than just one schema and therefore must work at the level of AIM entities. This is to say a STEP translator needs a wrapper for every AIM entity that needs to be supported regardless of the schema it belongs to. As a consequence at this level CAD systems do not distinguish between schemata and hence APs anymore.

A STEP implementer's true nightmare, however, arises from the fact that definitions of an AP may change dramatically by moving the AP from one stage of approval to the next. Such a change can be so drastic that it can jeopardize a production process where data exchange is based on STEP to the point where the production process could break down if those changes are not caught early enough.

A very good example is the move of AP203 from DIS to IS. Fortunately this happened at a time where very few data exchanges based on STEP were in use in industry. Two things of far reaching consequences were introduced. First of all, practically every AIM entity was provided with an additional id attribute which was placed at the beginning of the list of attributes. For example

#100 = CIRCLE(#90,4.5); (DIS) became #100 = CIRCLE('id',#90,4.5); (IS)

As a consequence a STEP translator based on AP203 DIS couldn't read anything at all from a STEP file based on AP203 IS without adding a little if-statement to skip the first attribute. But this had to be done to each access routine and every software engineer knows what it means: A tedious search in the whole source code and the nagging thought that one location might be missed. Needless to say that matters were complicated by the need to introduce a switch between the two versions of AP203 to support both for some time.

Even worse was the following change. There exists an AIM entity called SHAPE_DEFINITION_ REPRESENTATION which contains two reference attributes, one pointing to the geometry, the other to all nongeometrical information. In the very early days CAD system STEP translators were only looking for the geometrical information i.e. started processing a STEP file below the said AIM entity i.e. the problem didn't materialize immediately. But by the time some nongeometrical information needed to be processed as well, one had to start with the AIM entity SHAPE_DEFINITION_REPRESENTATION and surprisingly the geometry did not appear anymore. What has happened? Well, the order of the two reference entities was reversed going from DIS to IS i.e. STEP translators were still looking for geometry in the same place in IS as in DIS before but were given nongeometrical information.

To sum up it is very tedious to track all if statements to handle the various cases and to manage the constantly changing data schemata.

Last but not least, STEP is also difficult to explain to CAD users. A CAD vendor being present in both the US and European market must explain the existence of two APs (203 and 214) to do basically the same thing, namely the exchange of CAD geometry, and of course support them both.Once the CAD users have understood this they start asking questions like: Can I exchange Surfaces with AP203 ? Or: Can I exchange colors with AP214 Or in general: Are the two APs interoperable ? Well, the last question could be the subject of a lengthy scientific dispute I am afraid. Then the CAD vendor is pushed to further elaborate on his answers explaining that there exist conformance classes in each AP and what they really mean. The vendors need to explain the difference between topologically bounded vs. geometrically bounded geometry in case of surfaces or the fact that AP203 does not support colors while AP214 does and so forth. On top of this there is the issue of different APs being in different stages (CD,DIS,IS).

From a user's point of view great care is needed to pick the right conformance class to do the job at hand. The PICANT project at BMW needed at least AP214 CC2 to support the exchange of assemblies, which is not supported by CC1. Due to the complexity of the assemblies the desire was to split assemblies into smaller ones. As this could not be done by CC2 the PICANT project needed to use AP214 CC6 to accomplish this.

Recently the situation has gotten aggravated by the fact that AP203 shall be extended by modules to cover more functionality. Putting all this together leads to a whole bunch of settings needed for STEP exchange and it keeps the support engineer busy to explain and communicate the right ones for a particular pair of CAD systems.

So once again, though from a broader perspective, STEP is not necessarily implementation friendly.

To summarize, mastering and implementing STEP is ressourceful for a CAD system vendor. There are royalties to pay for third party software and travel expenses to follow up on the latest STEP developments. STEP binds at least 3-4 software developpers in the startup phase and at least one software engineer full time for later maintenance and continuation of development. Lots of testing is needed to ensure that changes may not break existing production processes.

Now, the aspect of testing and thus ensuring smooth data exchange processes in productive use takes us to the next statement, which is a very signifcant experience.

Statement 2: Testing STEP based translations is a complex and difficult matter

A commonly heard buzzword in the STEP community is "Conformance Testing". The idea is to take a populated part 21 STEP file and check it, e.g. to see if it is syntactically correct or whether none of the rules specified in a schema are violated. Though this idea has its merits and has been used in the various testing events at ProSTEP and PDES,Inc. still the tricky question is "Yes fine, but against what exactly do I test for conformance ? " Here are some possible examples to refine the question together with a possible objection.

Do I test against a particular AP? -- But no one has implemented a complete AP !
Do I test then against Units of Functionality or Conformance Classes within an AP ?
      -- That's better, but it doesn't take into account the various approval status !
Do I test against a particular schema ? -- OK, but what about those best practices !
Do I test then against best practices as well ? -- Hmmm, but they are not documented anywhere!

One example for such best practices are the following three agreements reached at the vendor round table on the exchange of colors
Do explicitly write out your default color
Do explicitly write out RGB values if your color code is not supported
Follow an explicit sequence when overriding colors (solid, then faces, then edges)

Now, at this point you may have ensured conformance of your part21 file to some criterion above but unfortunately this does not guarantee you that you will end up with a successful translation. So, why is it possible that the translators of two CAD-systems conform to some criteria, but you can not get the data across ? Here is a list of possible obstacles
CAD models are corrupt in the sending system to begin with
CAD systems A and B have conflicting philosophies on how they handle accuracy internally
CAD system A supports exclusively NURBS, system B NURBS and analytical geometry
CAD system A supports cylinder topology, while system B does not
CAD system A supports procedural offsets, while system B does not
CAD system A supports parametrics(history), while system B does not
CAD system A knows layers(groups), while system B does not

as a consequence you have to check each individual pair of translators to obtain a substantial assessment on the quality of data exchange. Of course the question now arises, what could be possible criteria to assess the success rate. First of all one could use the percentage of all faces successfully translated. While this criterion might still be sufficient for surface oriented CAD modelling, it is by no means adequate for the exchange of solids, because solids require 100% success in the translation of all their faces. Therefore the percentage of all solids translated without loss of faces is a much better criterion and in fact it is in standard use today (e.g. in the PICANT project at BMW like anywhere else in the automotive industry). In most cases it is also required to be able to compute the mass properties within an error margin of about 1% before and after translation. It seems that this is the standard widely accepted in industry today when we talk about the exchange of solids.
However, one can also establish even stricter criterions like "Can I apply geometric operations on the imported solid?" or in addition "Can I treat the imported solid as it were created in my own system (e.g. can I change blend radii)?". While in many cases geometric operations on an imported solid are likely to succeed if the accuracy philosophies are close enough, treating the imported solid as it were created in the receiving system is rather an ideal goal that is probably never achievable, given the fact that CAD systems differ in their modelling philosophies.
Besides that it is more important to ensure that in case of assemblies the correct positioning and structure (including sharing of parts) information is exchanged. Also, one would like to see the correct coloring or layer structure translated between CADsystems.
In the following subsection an account is given on how these ideas were put to practical use during the PICANT project at BMW in order to assess the quality of data exchange and how to identify areas of improvements.

PICANT: Experiences using STEP from a user perspective made at BMW

In 1997 it was decided at BMW to use the system Pro/Engineer in the process chain powertrain in addition to the already long existing CATIA environment. As a consequence a bidirectional exchange between Pro/Engineer and CATIA was needed to support DMU investigations and the socalled design in context. A typical example for a DMU investigation is the question whether the engine designed with Pro/Engineer fits into the front compartment designed with CATIA. In this case CAD data has to be translated from Pro/Engineer into CATIA. The other way round is needed for design in context where the electrical wiring done in CATIA needs to be imported into Pro/Engineer to check it against the engine. With these objectives stated the question arose which standard should be selected for the exchange of geometry. A thorough analysis was conducted and showed the following major results

VDA-FS could not be used because the exchange of assemblies was a must
For the exchange of the 2D portion of the model there was no alternative to IGES
Exchange of solids via IGES was impossible, but needed for e.g. the analysis of moments of inertia
The direct translation of CATIA solids to Pro/Engineer resulted in surface models only
STEP AP214 showed better results than the direct translation of solids from Pro/Engineer to CATIA using the criterion: Rate of solids exchanged without loss of faces
Without any optimization the success rate via STEP AP214 was about 50%

The conclusion was to use STEP AP214 but to also look for optimizations. First of all, faults in the implementation of STEP translators were identified and bug reports filed with the vendors. Accuracy related exchange failures were closely investigated and a harmonization of accuracies was found as a potential area of improvement. The proposal was to work with an absolute accuracy value of 0.012mm within Pro/Engineer and use a model space of 2000mm within CATIA.
Identified corrupt models were rectified using tools like the CATIA Healer/Optimator and the PS_STEPAdapter, both being products offered by ProSTEP.
Incompatibilities between the CAD modelling philosophies of Pro/Engineer and CATIA could partly be compensated for by establishing construction guidelines and "best practices".
Users were then trained to used these guidelines and also use system internal checks and not to ignore system messages that could indicate a later problem when exchanging the CAD model.
With all these measures except the fixes for the reported faults in place, BMW conducted a major data exchange test suite with 380 representative models. The rate of successfully exchanged solids could be increased to 80% and it seemed plausible that the rate could reach 90% with all the faults removed in the STEP translators, that had been reported to the vendors.

Statement 3: The services offered by ProSTEP and PDES,Inc. are invaluable
Among many services rendered to the STEP community at large by ProSTEP and PDES,Inc. the setup of socalled vendor round tables proved to be of invaluable help for fostering the harmonization and maturity of STEP translators thus enabling STEP based CAD data exchange to enter production processes at companies.
Even though the participating vendors are competing fiercely on the market, dedicated individuals from all the vendors were highly committed to improve their STEP translators by freely communicating information and willing to test STEP translator implementations in a prereleased stage. The round tables also proved to be an excellent forum to meet STEP experts face to face to discuss implementation versus standardization issues.
In essence the round tables were invaluable in detecting potential problems early on, which could be resolved before release. Many such resolutions have been formulated as best practices.
Another service provided by ProSTEP and PDES,Inc. has been the organization of socalled test rallies where a selected set of test cases were formulated and each pair of exchange evaluated to the benefit of

detecting all kinds of issues. While the test rallies were carried out on the basis of using prerelease versions of STEP translators, ProSTEP in particular also holds benchmarks on a regular basis with released STEP translators and evaluates their maturity and capabilities. The results are made available to the public.

Also the working group Quality and Test at ProSTEP, which has a long standing history of being chaired by a BMW Group representative, has always played a key role in helping STEP translators mature. Through close collaboration with the vendor round table vendors became aware of real existing everyday data exchange problems and most of all were provided with real production models so that the vendors could study the reasons for unsuccessful data exchanges.

The working group Quality and Test also serves as the driver for the ProSTEP benchmarks on released STEP processors.

4 Conclusion

After having been involved for almost four years in the development of STEP translators and witnessed the leaps forward it is the author's understanding that one may very well claim that STEP as a technology has reached maturity with respect to a couple of aspects. STEP is certainly superior to VDA-FS and IGES, particularly when it comes to the exchange of solids and assemblies. STEP translators today are sufficiently stable and mature for the productive exchange of 3D CAD data. The standard itself is per se no major roadblock anymore, because exchanges possibly fail for other reasons as has been explained with the experiences made within the PICANT project at BMW. Furthermore lots of optimization tools are available today for the STEP based exchange of CAD data.

Nevertheless there are challenges to master in the future. Probably the greatest issue is the fact that most parts of the standard are still not in an IS stage which means that changes are likely to occur that may negatively impact existing productive data exchange processes. Great care is needed and advisory.

On the other hand STEP translators still do not cover important areas like drawing or annotation. In addition STEP translators do not cover the support of CAD/PDM integration. To the author's knowledge some vendors have started working on drawing, some others on annotation and yet other vendors are addressing CAD/PDM integration in 1999. It shall be stressed that not everybody is currently working on the same subject contrary to the years before where the common goal was to make solids and assembly exchange work.

5 References

[1] www.prostep.de , www.stepnet.org
[2] www.rdrc.rpi.edu/express-v/homepage.html

Progress Towards Solving Common CAx Challenges in the Automotive Industry

Brian Shepherd

PTC, Waltham Massachussets, USA
bshep@ptc.com

Abstract: Today, the automotive industry faces many challenges in the productive use of their Cax toolsets. This paper outlines 8 problems and some progress being made against those problems by Parametric Technology Corporation. The problems are encountered while pursuing these common industry objectives:

- Creation of good quality shareable models.

- Easing the import and update of models between disparate systems.

- Enable true extended enterprise collaboration.

1 Design Synthesis

CAD systems today typically help users create and change geometry - not solve problems or meet design requirements. PTC is addressing this challenge through a technology called Behavioral Modeling. Behavioral Modeling allows users to add virtual instruments to their CAD models, called analysis features, that measure the performance of the design against requirements. These instruments can measure quantities native to the CAD system or connect to external analysis tools. Further, the user, through the use of sensitivity studies or Design of Experiments, can explore the design space. These techniques give the user a deeper understanding of their design. Finally, behavioral modeling includes the ability to drive to an ideal solution using optimization algorithms. As a package, behavioral modeling allows system designers to focus more on the objectives of the design and enables more innovative design through an automatic iteration goal seeking approach.

2 Connecting Simulation to the Design Process

While broadly used in automotive design, computer aided analysis is not really integrated into the mainstream design process. This leads to disconnects between what is analyzed and what is designed. Further, without integration to a data management system and to the behavioral modeling techniques discussed above, the location of the analysis models not to mention the rationale for the analysis is often lost. PTC has focused on this integration of CAE into the design engineering process through a set of simulation tools that are CAD geometry based, feature technologies that allow engineers to get better, more trustworthy results with less manual effort, and which are integrated with the PDM system. Recent improvements in the underlying CAD system allow for simplification or abstraction of very detailed CAD models to geometric models more suitable for numerical analysis. This is done without breaking the associative link with the fully detailed models so that changes in the design model can be quickly and automatically assimilated in the analysis model.

3 Using Massive Models

Today's CAD tools allow for the creation of very rich product models with a huge amount of both detail and embedded design intent. While important to have in many situations, this volume of data can be troublesome at times when lightweight models are required for increased compute hardware responsiveness or when sharing with parties outside of the direct enterprise. PTC has introduced "shrinkwrap" technology that, with some simple user guidance, creates a model that captures the outside geometry of a model while throwing away the internal details. The analysis properties of the model, such as mass and center of mass, are not changed. This provides a method for generating a lightweight version of a fully detailed model for use in the next higher level assembly. Further, the lack of internal geometry protects a company's intellectual property when sharing the model with customers or partners. PTC's technology can reduce model size by up to 90% while still providing accurate models for fit and interference checks.

4 Maintaining Model Quality

Today's parametric tools are both easy to use and easy to misuse. User's can create models very quickly, but those models may not adhere to either company standards or to recognized best practices for parametric modelers. Such models may look good but can wind up being very hard to change, share, use downstream for manufacturing or analysis, or reuse. PTC has begun to provide tools for quickly finding, and in some cases automatically correcting, common problems in CAD models. This interactive tool, analagous to a spell checker in a word processor, can help the user identify and correct mistakes as they are made. With today's CAD tools, it is always much more economical to correct a modeling error as it is made rather than months later.

5 Controlling Parametric Assemblies

Similar to the above point, the ability to create geometric references between components in the assembly models of modern CAD systems is a powerful feature for creating models with embedded design intent and which change in predictable and desireable ways. However, it is also easy to create very complex, undesired relationships between models if the user is not experienced or careful. This can lead to an assembly model that is difficult or unpredictable to change. Leading CAD tools must provide some tools and guidance to help users create "good" relationships and avoid "bad" relationships in their assembly models. PTC's top down design tools implement a concept called reference scope control. This allows a knowledgeable or lead engineer to define what kinds of relationships a user can create in the assembly. For instance the lead can define that components may reference only other components in the same level of subassembly, reference only a skeleton model, reference anything at all, or reference nothing at all. The user is provided visual feedback as to what components can and cannot be selected for assembly references. These techniques lead to well constructed assemblies that contain the user's design intent but which can be easily changed or reused.

6 Geometry Healing

In the automotive world of today, moving data between CAD systems is a fact of life. Often times, due to dissimilarities in the systems, geometry must be manually "repaired" before it is useful in the new system. Many CAD tools provide tools for moving vertices, trimming curves, extending surfaces, and managing topology to enable this manual repair. However, on a complex model, this manual process can take hours or days and is very exacting work. This slows down the design process and leads to user discontent. Through new objective driven healing technology, called Import Data Doctor by PTC, the time required to accomplish this cleanup can be greatly reduced. In this revised workflow, the user selects a portion of the imported model to work on. This simplifies the task into manageable chunks. Then the uses specifies the objectives of the cleanup. These objectives can be surface connections, tangency, and leader/follower relationships. Additionally, some surfaces can be selected for no change in order to preserve things like styling intent. Following specification of these objectives, the model can be automatically changed to meet these constraints. A preview option is available in case the result was not as intended. This workflow and technology allows the user to work at a higher level of abstraction on the cleanup problem leading to faster repairs and happier users.

7 Change Propagation

As automotive companies and suppliers must work with many different CAD systems, as described above, the need to propagate not just geometry but the changes in the geometry becomes vitally important to a productive design environment. PTC has delivered an Associative Topology Bus that allows dissimilar CAD systems to be connected together in a way that shares both geometry and changes in that geometry between systems. This allows disimilar design tools to be used effectively in a design process and moves data exhange from a one-time event to an on-going, behind the scenes, process inside a large organization.

8 Supply Chain Collaboration

As automotive OEM's continue to outsource more and more design work, the entire supply chain must be connected in a collaborative design environment. This includes the requirement to construct and share a comprehensive Bill of Materials no matter where the item masters are stored. It also mandates the ability to visualize data that is located anywhere and from any source so that progress reviews and multi-location decisions can be done quickly. PTC's Windchill product line uses the federated web server approach to enable the creation of coherent views of the product structure and the geometry regardless of who did the work or where.

STEP – a key element for the PDM strategy at BMW

Dietmar Trippner, Thomas Kiesewetter

BMW Group, Munich
Dietmar.trippner@bmw.de

1 Current situation and goals

Current and future demands in the vehicle development processes result in highly complex requirements to the underlying Cax- and pdm-technology. The growing complexity of the product in its parts leads to the necessity of a lot of functional and structural enhancements in a pdm system architecture. The handling of vehicle variants in very early stages of the product life cycle with a deep integration to the later logistics variant management is a typical requirement in today´s vehicle development processes. The goal is to achieve a complete digital representation of the car by all describing data elements and documents.

The "time to market" process can only be enhanced by a better integration of data, processes and systems as the basis of the pdm solution. In the result every user in the company involved in the development process shall be enabled to access the required data in the latest stage, in an user- or discipline-specifically optimized representation and with easy and fast access methods. The goal in view of the semantic and syntactic models to support such an environment is a close integration of the underlying product, process and project data models.

Boundary conditions in the hardware and software area in the field of pdm led to an existing pdm infrastructure at BMW which is similarly found at other car vendors too. A large number of "inhouse" solutions was developed when products were not sufficient or the mainframe systems were not supported by the pdm vendors.

Beside those company wide used own developed pdm solutions a large number of smaller systems was developed because of reasons as:

- fast availability
- very specific functional focus
- highly granular data models
- mapping to main systems not possible or to expensive.

In the following the pdm situation at BMW is described related to a part of the whole pdm focus which is described in picture 1. The main focus of the CAD engineer today is not the single part but the total vehicle or hugh parts of it such as the front or specific modules. The complete assembly from the geometric, functional and production technological view becomes of great importance to check all requirements as early as possible. The increase of CAD data amount at BMW grows strongly as well as the number of data exchanges with suppliers necessary to support external modeling. The result of a world wide usage of BMW´s inhouse system PRISMA for CAD data management as well as part management is the necessity to have this system up and running with performance 7x24 hours a week. The number of regular system users grew to over 6000. Every PRISMA related development becomes an "open heart operation" which has to be handled with extrem care.

In several departments very specific pdm requirements grew with respect to manage f.e.:

- 2D and 3D characteristics
- simulation results
- chassis specific data
- specification documents
- electronic components data
- control module information
- measurement results.

136   D. Trippner, Th. Kiesewetter

Vision
Product Data Management

Source: CIMdata

Picture 1: Total scope of pdm

When these requirements could not be fulfilled by the central systems in the right time frame the decision in departments often was to develop own solutions which were expensive to realize, hard to maintain and inflexible for future extensions. In addition many of those solutions needed interfaces to those pdm solutions which comprise BMW's pdm backbone. Areas where departmental or process specific solutions were implemented are described in picture 2.

A well integrated process driven total pdm solution addressing requirements as:

- data consistency
- data integrity
- configuration in specific views
- management of variants and alternatives
- application integration
- business process to workflow mapping
- new architecture
- scalability and others

needed a redefined future oriented pdm strategy which all involved departments and users agreed to.

2 Approach

The summary of current issues for pdm in the development process at BMW leads to four key aspects:

- Increasing usage of PRISMA leads to necessity to enhance its functionality while keeping performance.
- The number of "local pdm solutions" grows and leads to non-well-integrated data and processes.
- Data has to be deeper integrated within the development process and through to manufacture and sales.
- The scalable pdm integration of suppliers plays an important role.

CA PDM Landscape at BMW 1999

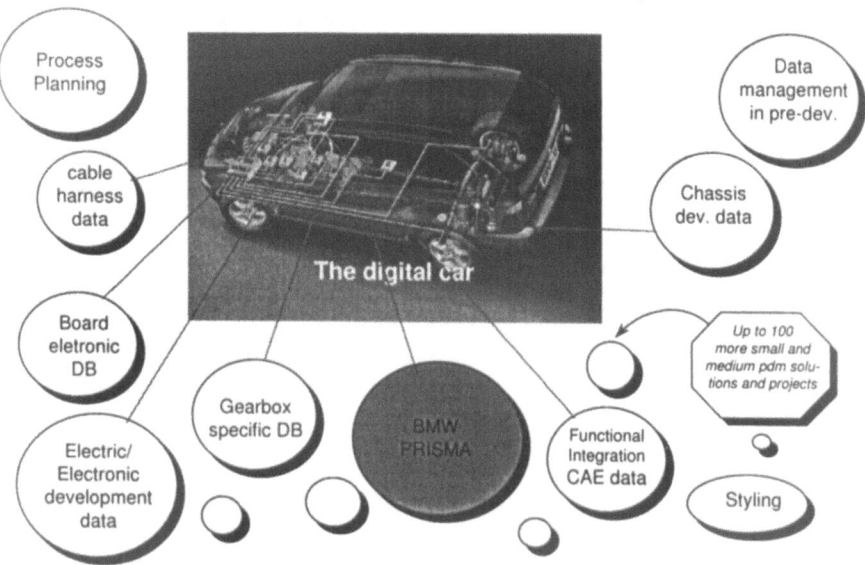

When the pep-pdm strategy (pep = product development process) was elaborated it became clear that a one-system-solution covering all data objects, processes and application areas was measured to be unrealistic. Different pdm projects are known where this approach led to several years of conceptual work, late implementation with a negative result at the end. This approach is hard to handle and from a system´s perspective nearly impossible to be supported in a sufficient and performant manner.

A multi-system solution leads to the necessity to build up an innovative integration and communication framework to support functional and flexible integrations. One of the main advantages besides the fact not to be bound to only one vendor is to be flexible for specific areas. A global multi level pdm architecture with several systems in place can only be handled if the number of systems is kept on the absolute minimum level.

Basic strategic elements build up the pep pdm strategy:
- **Standards and standard software where possible**
  - Reducing the number of different systems
  - Openness, flexibility and system integratability
  - Support of standards (STEP, CORBA; PDM enabler)
- **Increase of integration**
  - CAx (Integrate geometric items and relations with PDM data)
  - "Local" PDM solutions for technical data management
  - Virtual and hardware oriented processes
  - Support of the BMW PDM backbone concept
- **Increase of functionality**
  - Expanding product data model and PDM services
  - Product configuration (supporting the early phases of development)
  - Workflow (company wide vs. department specific)
  - Extranet (PDM integration of suppliers)

Whereever possible standard software has to be used instead of "inhouse" implementations. Before that decision migration aspects as cost-benefit-analysis as well as differentiation possibilities from competitors have to be taken into account. Of greatest importance for systems to fit into the architecture are their openness, flexibility and integratability. Only the commitment to standards gives the chance to deeply integrate different system classes where STEP, CORBA and the PDM Enabler can be mentioned. The specific role of STEP is addressed in the next chapter.

The deeper integration of CAD data management with part data and other specific information is a critical success factor for the processes of geometric integration (GI), functional integration (FI) and productional integration (PTI). Where necessary technical data management solutions are accepted when data is not stored redundantly, a requirement which can be fulfilled by referencing instead of copying information. Even if the vision "Digital Car" is realized in several aspects there are different situations during the vehicle development where physical mock-ups are still necessary. The hardware-oriented processes have to be managed in an integrated manner with the virtual processes by an integrated pdm solution, picture 3.

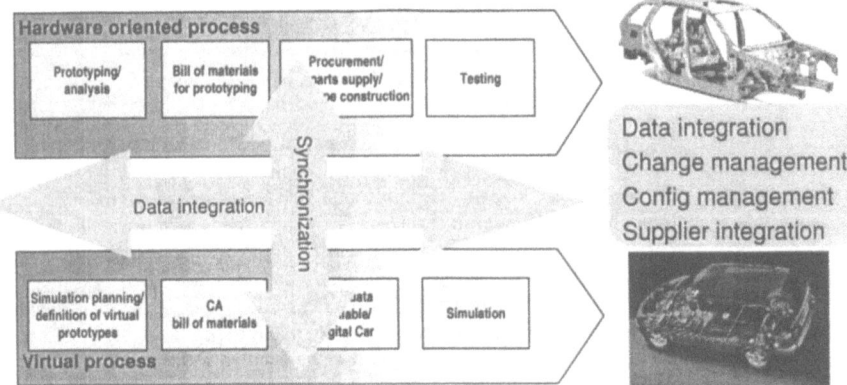

Picture 3:

The reference concept that all solutions have to follow is the BMW pdm backbone concept, picture 4. Several steps are necessary to increase the resulting pdm functionality, where the product model expansion and the extension of the pdm services are a first part. A consistant product configuration from the very early stage to the later logistic stage with more than 10^{30} variants is the required next step. Company wide central workflows have to be synchronized with most flexible development workflows and lifecycles. The supplier integration asks for a solution scalable between highly functional design-in-context and WEB-based metadata-access.

PDM Backbone concept

Definition:

At BMW, the term **product and process data management backbone (PDM backbone)** is used to refer to the collection, management and allocation of product and process description data which is of global importance and for which clear and binding general guidelines, responsibilities, processes and system architecture must therefore be defined.

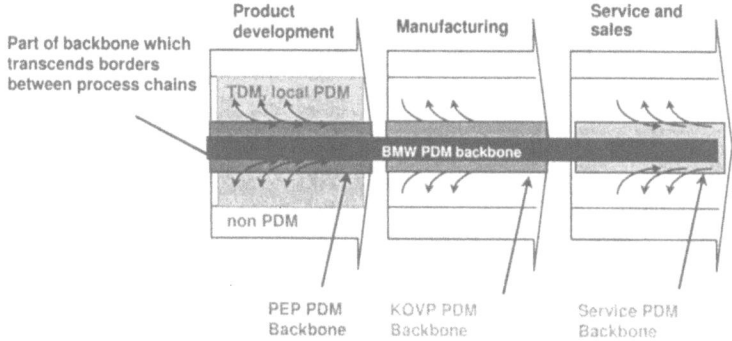

Picture 4: The pdm backbone concept (logical view)

The first necessary action to realize a strategic integration of several locally home grown systems based on Microsoft Access or OracleWeb into a consistant, redundance free global solution was to build those on the same platform. This one therefore has to be most flexible, well integratable and distributable. The demand of rapid implementation as well as a long list of available functionalities led to the selection of the InformationManager (iMAN, UGS) during a pdm-toolset-benchmark.

At BMW all solutions based on iMAN will work on top of a commonly customized BMW specific object model, use the same in iMAN integrated interfaces to backbone systems and work with the same workflow templates. For implementation and user-interfaces guidelines were already defined which have to be followed. Every customization follows the goal of minimum customization to ensure the avoidance of over-customization. Each single object or function implemented is managed by a central software management tool to realize maximum synergy effects.

Only complementary data to that stored in the pdm backbone is allowed to be handled by iMAN solutions inside BMW.

The first step in the new pdm strategy has been taken and is currently under successful work. Beside necessary enhancements of PRISMA the next important step will be the evaluation process for the future pep pdm solution. Here the usage of a commercial pdm system is also planned and first work on architectural design and functional as well as technical target catalogue is done. The focus on standards to realize the necessary integration in view of data, functions and processes is a strong must to ensure openness of the solution, picture 5. For the part of data integration as well as the definition of enterprise wide data models the standard STEP AP 214 and the pdm schema are a good basis to build on, [1]. International teams worked out these data models which play a significant role in the data exchange area.

Goal: Integration based on standards

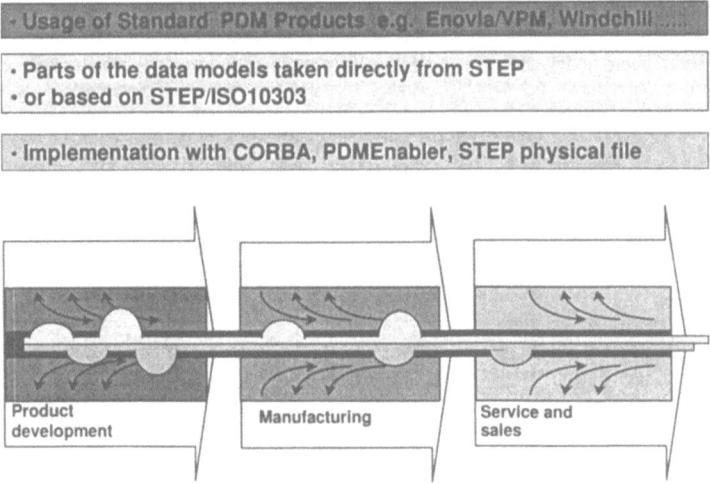

· Usage of Standard PDM Products e.g. Enovia/VPM, Windchill .....

· Parts of the data models taken directly from STEP
· or based on STEP/ISO10303

· Implementation with CORBA, PDMEnabler, STEP physical file

Product development | Manufacturing | Service and sales

Picture 5: Utilization of standards and standard products to build pdm solutions

In order not to start data modeling directly in pdm projects from scratch it is sufficient to follow the STEP AP 214 where possible and useful. By doing so this standards can be seen as a reference. Projects utilizing STEP at BMW are described in the following chapter.

3 Role of STEP

The goal to reach a well integrated and harmonized multi-level pdm solution corresponding to the defined architecture can only be reached if the role of standards for data definition as well as for communication is pointed out clearly. At BMW a team is working on the integration of existing object models and is responsible for a central object glossary and data dictionary. For the object model description the OMG standard Unified Modeling Language (UML) is used. UML in addition to object modeling allows for the definition of use-cases to define application scenarios and their correlation to data entities.

The communication standard which is focused at is the Common Object Request Broker Architecture (CORBA) which is highly supported by new versions of pdm solutions on the market. For those technical data management solutions realized at BMW using iMAN in an distributed environment (D-iMAN) a commonly used basic object model has been worked out which supports interfacing with legacy systems. This one was developed in close reference to the AP214 as shown in picture 6.

It can easily be seen that iMAN itself is supporting STEP AP 214.
STEP at BMW plays an important role in the following projects and activities:

- Development of the enterprise data model for product development
- Pdm platform development PPEA (iMAN)
- Integration of PRISMA and TAIS
- Integration of Pro/E and CATIA for DMU etc. (based on Pro/IL-Prisma integration)
- Data exchange with suppliers

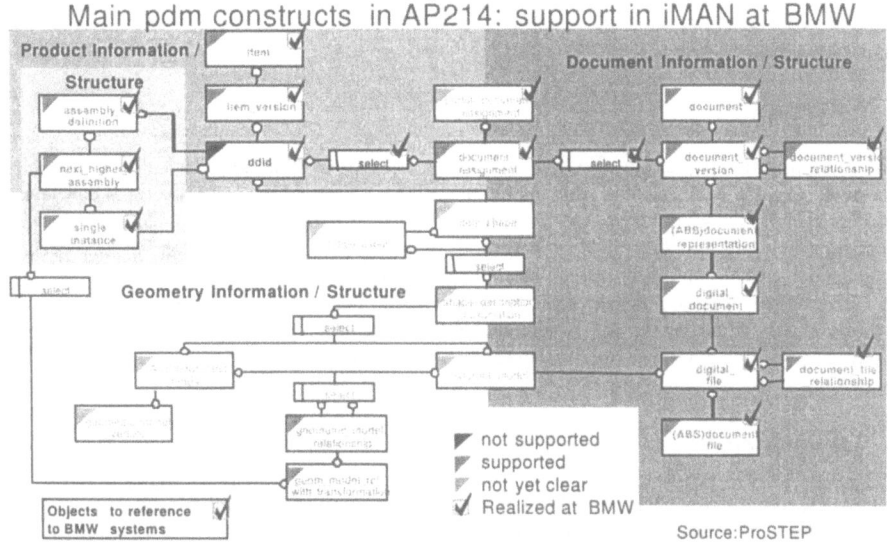

Picture 6: Main objects of the PPEA solution (PDM Platform for Engineering Applications)

During the initiation of the project PICANT (Pro/IL-PRISMA integration) the mapping of main objects handled by the involved systems with the STEP ARM schema as well as the overlapping of schemas has been evaluated. Some of the results are shown in picture 7.

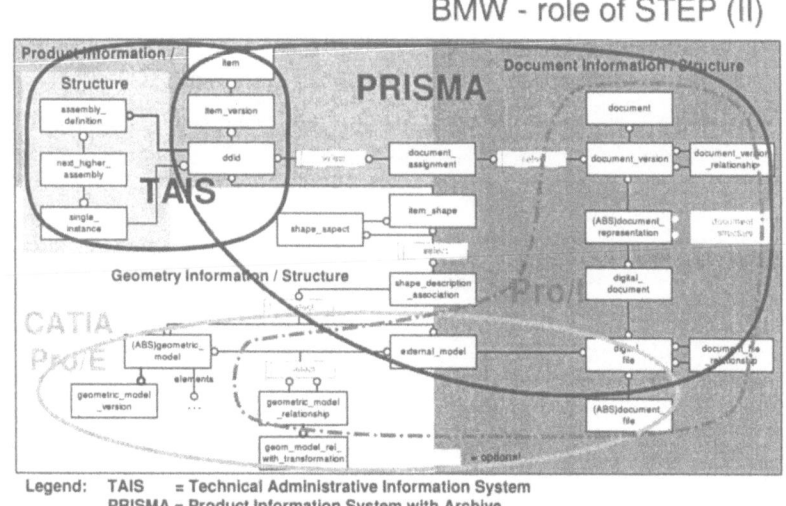

Picture 7: Role of STEP for PRISMA-TAIS-Pro/IL

The relevance of STEP, especially AP 214 and the pdm schema, as a reference for data modeling projects at BMW ist still growing. BMW is participating on ProSTEP's pdm round table and is member of the ProSTEP e.V. On a regular basis STEP experts of ProSTEP GmbH support BMW projects in the initiation and implementation phase.

4 Summary

The standardization process around pdm and related topics is a strong requirement to be capable of managing world wide distributed pdm environments [2]. To keep in track with the rapid development in commercial pdm solutions as well as in the cad arena it is of great importance to speed up the standardization process. One right step towards this goal is the establishment of ISO technical specifications which could be released in less than one year.

The influence of STEP to new pdm solutions on the market can easily be seen in object models which are very closely related to the AP214 reference model. The exchange of structures, metadata and f.e. cad data can be supported by easily implementable pre- and post-processors.

5 References

[1] Anderl, Trippner: STEP Eine Einführung in die Entwicklung, Implementierung und industrielle Nutzung der Normenreihe ISO 10303 (STEP)

[2] Spur, Krause: das virtuelle Produkt – Management von CAD-Technik / Hanser

'TOGO' - The New Integrated CAD/CAM/CAE System at Toyota

Shinji Yuasa

Toyota Motor Corporation, Information Systems Development Div., Aichi, Japan
yuasa@mail.toyota.co.jp

Abstract: As the competition for new car development accelerates in the auto industry, we found a truly integrated CAx system was crucial to achieve concurrent engineering and digital master that would contribute to shortening car development cycle and reducing manpower. "TOGO", which means "integrated" in Japanese, was developed as a solution to this situation. It is equipped with the edge technology such as G2-continuous styling surface modeler, solid modeler and parametric modification engine. On top of them lie a whole range of application-specific functionalities that enhance the productivity throughout the automobile development processes. Today the deployment of "TOGO" has finished within the company. During its development and deployment, we have gained numerable important experiences concerning the system architecture as well as end user relationship. We are now shifting our focus to digital mockup, engineering data management, and engineering know-how database.

1 Background

It is an extremely important strategic task for any automobile companies to shorten the development cycle of new cars and reduce the manpower for it. Toyota has been devotedly engaged in this challenge through the use of computers since 1960's. Fig. 1 shows the conventional CAx systems for styling design, body structure design, stamping die manufacturing, and so on. Unfortunately, although these systems helped us to reduce the man-hours required in each stage of the development cycle, they did not contribute much to the compression of the entire cycle because of the following limitations :

1) Heterogeneous and low-quality data
   This complex system configuration was the principal hurdle for concurrent engineering, because each time the data file went through a stage, it had to be converted, part of information was lost, and delay was incurred. Moreover, feedback from later stages to upstream was not easy because data transfer mechanism was not complete for this direction.

2) Limited 3D Modeling Capability
   Three dimensional design was not sufficient especially in the domains of powertrain and chassis due to the limited capability of the CAD modeler.

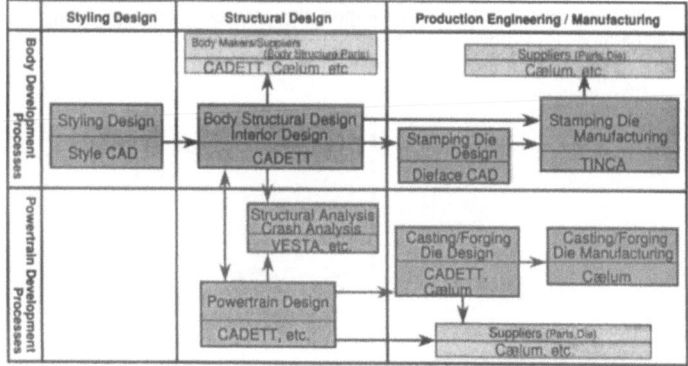

Fig.1 Conventional Cax Systems at Toyota

2 The Concepts of TOGO system

In 1991, Toyota decided to replace all of the old systems with a new integrated one which we named 'TOGO' after its concept. Fig.2 shows the TOGO architecture, whose platform is based on an commercial CAD/CAM system CADCEUS which is developped by Nihon Unisys. Also, we adopted STRIM (developped by Cisigraph) geometry library for styling application. The basic ideas of TOGO are as follows.

1) Integrated system and database

First, all the applications necessary for the development of automobiles are integrated into one system and a single database chain runs end to end through the mainstream processes. In other words, the idea is to allow all the people concerned to share the same geometric data which is the core of engineering information, without the hassle of data conversion or compensation for inconsistent data accuracy, and let the data go back and forth between up and downstream processes without loss of information.

2) Application specific functionalities and user-interface

The second concept is to provide each development process the most suitable application for the job. While sharing the database among processes, applications need to be specialized to accommodate themselves to design methodologies and operational improvements at various stages such as styling design and body engineering, thus speeding up the total development time.

3) Hybrid Modeler

The third concept is to build a hybrid system which has both surface and solid modeling capabilities. Since TOGO's architecture is rewritten from the scratch, wireframe, surface, and solid models are equally handled by its geometry engine. A surface can be seen as a conventional simple surface or an open solid. A solid can be treated as an ordinary solid or as a collection of surfaces each of which the user can apply surface-oriented operations to. Above this geometry engine lies a parametrics engine. It parameterizes everything by remembering the history of any operations to any entities. Not only ordinary solid operations like sweeping a surface to define a solid or merging two solids can be parametric, but a combination of surface- and solid-oriented operations can be parametric, too. There is no seam between parametric solid and parametric surface. On top of it, custom applications are built. They maintain conventional operability and functionality, while offering the power of newer technology like solid and parametrics.

4) Assembly-based collaboration support

In Fig.3, the topmost assembly is owned by a supervisory engineer and the lower three constituent parts are separately designed by three different engineers. This is not a simple check-in / check-out mechanism. The assembly and the parts are parametrically linked so that a change in one file propagates to the others and triggers parametric deformation in them. All of this linkage occurs among separate files. Everyone can control when to import the parametric change. Time paradox doesn't occur because the parametrics history remembers the timing of reference.

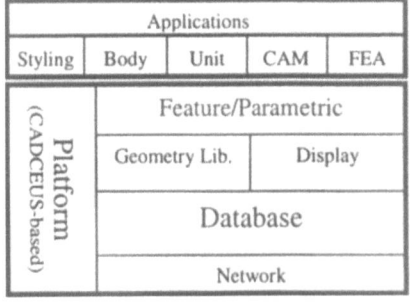

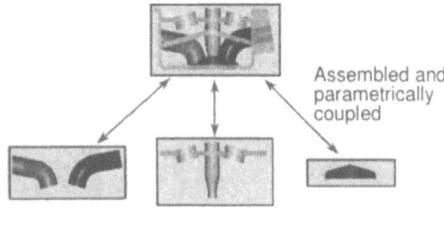

Fig.2 TOGO Architecture   Fig.3 Assembly-based Collaboration Support

3 TOGO applications

Fig.4 shows the range of TOGO applications.

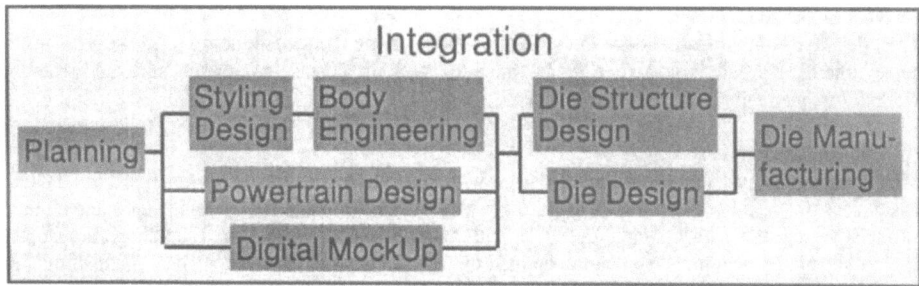

Fig.4 The Range of TOGO Applications

(1) Styling (Fig.6-7)
For styling design, thanks to STRIM technology, we can have very smooth surface continuity across the boundary of adjacent surfaces. They guarantee G2-continuity which means continuous curvature distribution.
Because the surfaces are quite smooth, they are ready for direct NC machining without having to smooth them or resurface unlike we did in the past.
We also have a U.S. patented rendering algorithm that calculates extremely accurate color. We plan to use this technology for model-less data-only design approval and CG catalog making.

(2) Body Sturcutre design (Fig.8)
For the body engineering domain, we have a set of conventional wireframe and surface modeling functions with enhancements in surface quality and connectivity. TOGO supports functions specialized for automobile design such as window shield design and evaluation of visibility through a mirror.

(3) Wire-harness design (Fig.9-10)
TOGO is equipped with functions for both cable routing and circuit wiring for wire harness design. This is an example of cable routing around the engine block and this is a circuit diagram drawn by TOGO. The wire harness system of a car is destined to have many variations due to the large number of optional equipments. TOGO can handle multiple variations of circuits in a single data file, from which it can generate connector and cabling diagrams for each variation.

(4) Powertrain design (Fig.11)
In the arena of powertrain design, TOGO's fully parametric solid modeler fits well. Fig.11 shows a complete engine block with fillets and drafts modeled with TOGO. Since this model is parametric, most of the major dimensions can be modified. To build this large model, surface-oriented functions are also required. In this domain, assembly and collaboration support which mentioned in Chapter 2 is welcomed by our engine designers.

(5) FEA (Fig.12)
TOGO's FEA module has both pre-prossessing and post-prossessing capabilities. Especially, our pre-prossessor is unique in that it can generate meshes for not only a surface model but also a wireframe model. This allows us to analyze the design very early on when the model is still a wireframe one, and to feed back the result to the design before it is a complete surface model. Of course, automatically generated meshes can be optimized by hand. Attributes like thickness and material are retrieved from our BOM system to eliminate the trouble of manual input. Altogether, a fairy accurate result out of a small number of meshes with a medium supercomputing power has been realized.

(5) Digital Mockup (Fig.13)
Since early design is partially wireframe-based, we do layout design in two phases.
First, layout planning and checking are done on TOGO using an interference and clearance checking function which works fine with even wireframe models. The result is fed back to each responsible designer on the spot.
Then detailers come in and surface the wireframe model using the auto-skinning function. This takes some time. The result is sent to a digital mock-up tool for concise evaluation such as assembly workability, maintainability, and appearance.

(6) Die-face design (Fig.14)
Most commercial systems cover dieface design with general-purpose modeling functions, but TOGO has dozens of specialized commands and a framework for this domain. Addendum shapes and dieface surfaces are created with a simple operation. Designers can analyze how the final shape is mapped back to the blank sheet metal so that the optimal dieface shapes and bead arrangement are determined. Stamping direction can easily be adjusted. A single data file can represent multiple process models such as draw, bend, and so forth.
The output from dieface design is sent to an FEA tool to evaluate formability, and then to the NC module for CL generation.

(7) NC (Fig.15)
NC machining requires a broad range of experience and know-how. TOGO keeps this information in terms of a know-how database, which then selects the most suitable machining conditions for a given work.
Now that we have almost finished the development of TOGO. As shown in the examples, TOGO system has been deployed widely through new cars development process. We had finished the replacement of the old systems in body development process by 1997, and we have completed the replacement of all old systems in unit development process this year.
As a result, in 1998, we have more new car development projects in shorter period, while the number of engineers keeps still. We believe that TOGO system contributes lots for this great progress.

4 Further challenging items of TOGO

(1) More engineering support
Firstly, we are linking TOGO with other systems (like BOM and data distribution system) through PDM and WEB technologies.
Also, in order to support more engineering work, we not only enrich our own functionalities, but also integrate domain-specific commercial components. Nihon Unisys and Toyota has collaboratively prepared OpenAPI for TOGO and Cadceus.

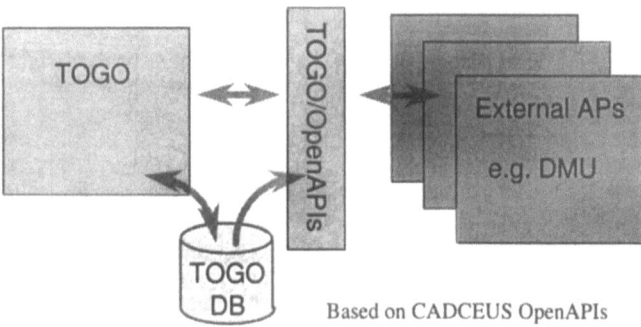

Fig.5 TOGO OpenAPI

Fig.5 is the diagram of TOGO OpenAPI approach. Through the OpenAPI, external applications can access geometry data more dynamically than via intermediate files. We would like to provide our engineers not only modeling tool but also real engineering tools in shorter period.

(2) Repeatable Digital Mockup

Digital mockup in an early stage of the vehicle design digital mockup capabilities should be applied repeatedly, especially in an early stage of the vehicle design. After the CAD data are approved and released, the management rule is strictly defined, e.g. engineering change procedure. So, at this stage, PDM system fits well. On the contrary, at the early stages of vehicle design, the model and configuration are dynamically changed. And the procedure is locally optimized in each division or module. From the engineers' or designers' point of view, they are working with several alternatives, and do not want to spend extra time for data clean-up. From the digital assembly engineers' point of view, the latest data without extra geometries are desired. We think that it is rather easy to provide PDM and DMU tools. However, it is much difficult to establish the rules as the base of the tools.

Also, we think that these tools and rules should be applied under distributed environments. For example, design is progressed in Japan, while digital assembly consideration should be done in Europe. Furthermore, suppliers should be involved in these simulataneus works.

(3) Knowledge Management

Engineers are working lots of pre-requisites or check items, such as regulations, production requirements, past problems, and so on. Also, engineers in charge of the module are often changed. It is rather difficult for them to understand what they should check at each design stage.

Current CAx systems can manage "What" information, however, "How" information still remains in each engineer's brain. Also, the old knowledge is not useful at all. Therefore, the knowledge should be dynamically updated, so called "adaptive". We believe that the computer support for capturing knowledge and navigating engineers is a next big challenging item.

5 Conclusion

The background, concepts, and application examples are introduced, and the further challenging items are mentioned , based on our TOGO development and deployment experiences. Through these activities, we believe that CAx systems will be innovated so as to contribute much to develop more attractive and ecology vehicle in a shorter period.

Fig. 6 Styling Design   Fig. 7 Rendering

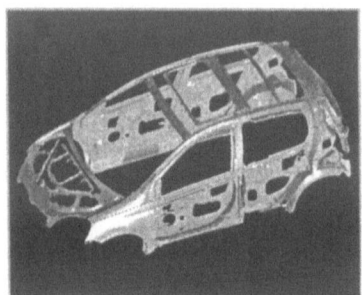

Fig.8 Body Structure

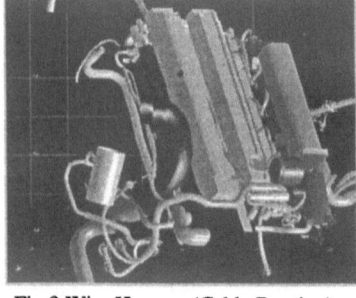

Fig.9 Wire Harness (Cable Routing)

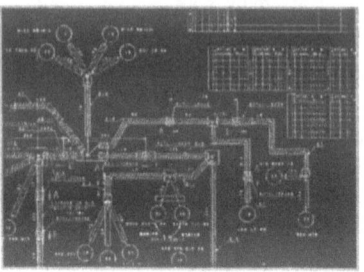

Fig.10 Wire Harness (Circuit)

Fig.11 Powertrain

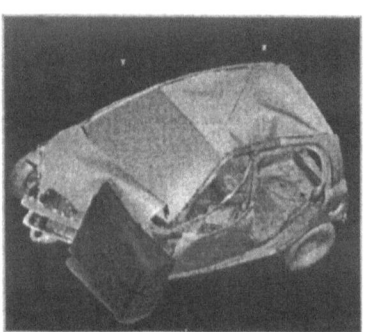

Fig.12 FEA (Crash)

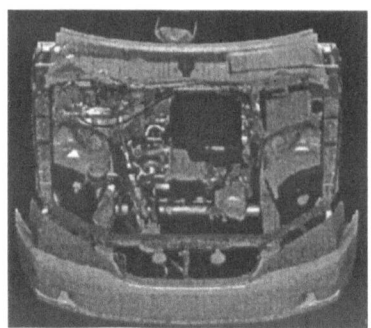

Fig.13 Digital MockUp

Fig.14 DieFace Design

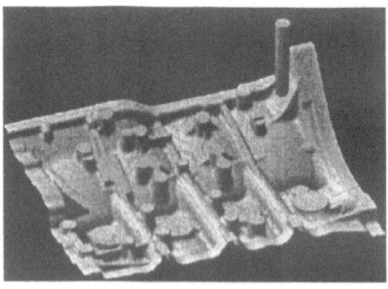

Fig.15 NC

Summary of Panel discussion

Barbi Driedger-Marschall

University of Kaiserslautern

1 Introduction

The panel discussion was lead by George Allen (Unigraphics), Dr. Joachim Betz (IBM), Prof. Werner Dankwort (University of Kaiserslautern), Prof. Peter Kellner (Volkswagen), Dr. Heinz-Gerd Lehnhoff (Opel/GM), Mr. Akihiko Ohtaka (Unisys) and Dr. Rainer Stark (Ford).
Prof. Dankwort as head of the board opened the panel discussion and formulated its objectives: The discussion was not only to summarise the past workshop days but also to offer the participants the opportunity for a brainstorming focussing on the topics "present needs in CAx" and "visions for 2007".

It is the nature of a panel discussion that a lot of current problems and possible solutions are be discussed. The summary report of the discussion cannot give a comprehensive concept, there will be contradictions as well as repetitions. The reporter tried to capture the most important ideas and the visionary atmosphere. It was surprising that the panel actually had no really conflicting opinions.

2 Brief Summary Of Workshop – Key Points

2.1 The Product Data Model (PDM)

The workshop has made it clear that the Product Data Model – Dr. Gus Olling from Daimler/Chrysler called it "Product Information Model" - is a big issue in current CAx.

2.2 "Dr. Gus Ollings' law and Dr. Lehnhoff's circle"

Concerning the emphasis put on certain factors of CAx, two only slightly differing views – both from representatives of automotive companies - were presented: The first version was explained by Dr. Olling, the other by Dr. Lehnhoff.
Dr. Olling believes that of all factors one has to take into account, 80% are persons (= human factor) and organisation, while technology is only 20% relevant – a distribution which Dr. Shepherd called "Gus Ollings' law".
Dr. Lehnhoff's model is rather a circle of which technology, the human factor and processes each take up 1/3.
Although these two versions are different, they show the same tendency, namely that - even though CAx depends on innovative technology - the human factor is of considerable importance when smooth efficient processes want to be achieved.

0 of 1

3   What Are The Present Needs in CAx?

3.1   Consultants

Technology has been the big issue of the workshop, the topic of consultants was missing although it is of importance. There is a need for skilled service people in the field of software products. This supports "Dr. Gus Ollings' law", because consulting is part of the human factor which – in Dr. Ollings' model – gets 80% significance. Only 20% are technology, and technology is software products, and the workshop has shown that the CAx software products are all the same (UG, CATIA, Pro/Engineer,
I-DEAS, CADCEUS etc.) and hardly any severe differences can be traced. Therefore the focus should be on consulting and service people and not so much on software technology.

3.2   Seamless Integration

The next speakers put an emphasis on integration. First, seamless integration is the topic. Function itself of software does not sell as major incentive, it is simply expected by most people. What will sell is integration, especially seamless integration, which is also the vision for the midterm future. The chief concern is: How can we achieve seamless integration? For the next years, the focus will be on having this kind of integration in data management. There are different approaches, but maybe there will be one PDM system and we will want to link an extended enterprise to it. In 2007, there might already be e.g. an agent technology so that we can link an application to such an environment. That will be one of the key technology and scientific research topics for the next years. Even feature technology will need this integration before it will be successfully integrated.

In the further course of the discussion it became clear that seamless integration is probably *the* big concern. The main goal is to do everything during the development process with CAx systems, to be able to run the whole process on the computer – in short: the digital product. Therefore, a lot has to be done to the functionality of CAx systems. In the next 5 years, CAD systems must be redesigned so that the designers and analysts will have proper tools and the PDM problem has to be solved. Presently there are problems with integrating TDM systems with CAD systems and different in-house data management systems, but this seamless integration is needed e.g. to achieve the vision of the digital car in the automotive industry within the next 5 or 10 years. One solution is the open system architecture, but it is not yet certain who will develop, implement and standardise it. Integration is needed for survival.

A lot has to be done concerning manufacturing engineering. With product engineering/CAD systems we already are where we want to be (except for the early design phase). We should nevertheless continue the good way we are on. But e.g. in CAM, we are far away from where we want to be with process engineering. We will not have seamless integration soon, it will take longer than another two or three years. The kernel-to-kernel approach would help, but e.g. the automotive industry should not be interested in how software suppliers achieve this goal. The automotive industry should instead be interested in breaking their engagement in software development in the CAx field to zero. This is not their core business. They can give advice, specify their technical and strategic requirements but they should no longer need to act as developers. The customisation effort should be significantly reduced.

3.3   100% Data Exchange Without Loss

Integration of design features and manufacturing features is more important than just design feature or software agent technology in the design area. There is not one single system which is very strong for every application domain or purpose. It would be easier to use only one system, but if application companies select the most appropriate system, we are forced to use two or more systems. In that case, the major challenge is how to exchange semantics between different CAx systems in a 100% correct way. STEP does not realise that, it is mainly focussing on the design area. We should concentrate more on the manufacturing area. The key concern is how to limit the road from early design to manufacturing with two or more systems without loosing any semantics. Long term archival of CAx data is another unsolved topic in this context.

3.4 Standardise Feature Semantics, Parametrics and APIs

When talking about seamless integration and 100% data exchange without loss, it has to be made clear what kind of date should be exchanged. When it comes to exchanging geometric information, either STEP or the use of a common kernel is a reasonable way. Exchanging feature information and knowledge however is an extremely difficult topic. It is a mistake to base engineering-in-manufacturing strategies on the assumption that this problem will soon be solved. Consequently, the kind if data that should be exchanged has to be defined in the first place, then we should think about the techniques to do it. We need integrated CAE tools and feedback loops to achieve optimal resp. improved designs. But: One of the barriers to this is the fact that many of these CAE programmes are used by very small companies with very specialised expertise. Currently, the costs for integrating these programmes with the major CAD systems is quite high. There are file-based approaches that work sometimes and API-based approaches that do not work so well. We should therefore focus on standardising CAx-APIs, this is a balance that needs to be fixed. This would also improve the freedom of choice between different systems.

Today, the definition of parametrics is very vague. It consists of many fundamental technologies, one of which is 2D-geometry. This is not ambiguous because it is mathematical. But the major CAD system parametrics is history-based. The design intent could be included in that history but it is not obvious, only very implicit. Do we want to exchange vague data? We should really define what kind of data we want to exchange. But: How do we define it? This leads back to the problem: How to standardise CAD system functionality.

3.5 "Future-Based" Design

The customer who will later buy the car should be on the mind of each designer. What are the requirements for this: legal, spatial or essential geometric, nice outer appearance. The question remains: How can the CAD system help to capture these requirements except as a string of text?

3.6 Rigid Personnel Structures

The culture of people concerned is very important. In Germany, employees stay in companies for a long time. It is necessary to enforce their training, update their technological know-how and refresh their knowledge. Otherwise they will stay in the technology they have learned 35 years ago. People should be trained in-house, not only by outside consultants. Within three years, all people concerned should be aware of what the company wants to achieve in technology and strive to reach this goal. Everybody should work with their individual strength. This might ask for a re-organisation in the company, but it will be more efficient. Processes and the organisation have to be adapted to the existing technology, they have to brought inline or they will be irritating. We need a seamless integration of tools, education and coaching of people from in-house (maybe with the help of outside consultants). Processes have to be brought inline with technology and people.

The way employees are educated has a large impact on their work at the company. In Germany, many engineers are trained at Universities, but the tools are still too complicated for the people who need to use them. Hence the tools must change. CAD tools must allow a more intuitive kind of learning. But: With knowledge-based design, for example, there are a lot of new things ahead. If we concentrate on existing tools, we are taking a step back.

In the European designer community, 80-90% are educated engineers. In the U.S. automotive industry, 80% of designers are not really engineers, they are mostly high-school graduates or have had a short apprenticeship. They get acquainted with the systems very easily. This leads to the conclusion that the

specialised, well-trained people should work in other fields, maybe as project co-ordinators. In Germany, there are very strict personnel structures and we are suffering from this inflexibility.
Becoming familiar with a new CAD system very quickly does not necessarily mean that this person is also a creative designer or good at finding new solutions. People should be taught how to support engineering creativity.

3.7   Intuitive Design / Feature-Based Parametric

From the software vendors' point of view, there has been a lot of talk about making design more intuitive, raising the level of design to the level of engineering thinking as opposed to geometric thinking. The CAD software vendors want to help achieve this goal. But designing intelligent feature-based parametric models is very difficult. Even if the usage of the design software can be made more intuitive, only few people will be able to do it.
Feature-based parametrics strive towards isolation because the stylist works on his own model and nobody else will be able to interpret this stylist's features. We have to reach a broader basis on which other people can also work. People also have to be able to interpret the results of engineering. There has to be a balance of engineers and data processing modellers.

4   The Participants' Vision for 2007

4.1   Growing Significance And Integration Of Suppliers

The degree of supplier integration will grow significantly. Different kernels will be able to communicate. We need protocols to transfer the data without loss, and for this we need the integration of suppliers.
There will be a stronger integration of suppliers and OEMs. There will still be at least four or five large software companies dealing with CAD and a lot of small software suppliers. Given suppliers will always sell the same software components to multiple vendors. Small CAx companies with a certain expertise will become suppliers to OEMs. Maybe these OEMs will become little more than software assemblers. The integration via the Internet will become even more important than today.

4.2   More Strategic Alliances

The number of strategic alliances between different realms as well as some other big mergers like DaimlerChrysler might come. This development will be enforced by people concentrating on strategic product lines. In the future, this will not always go along with a 100% system adoption. Such companies will have to find ways to interoperate with each other.

4.3   Increasing Significance Of Computers

It is hard to do strategic planning for information technologies for seven years ahead. Usually this is done for two or three years maximum. This technology develops so fast, it is hardly predictable. But of one thing one can be certain: Computers will be faster, so that a greater proportion of work will be done by computers rather than by human beings. People will still be necessary to do the creative things, but there will e.g. be a lot more CAE, a lot more simulation of physical events and processes. This simulation will be done at an earlier stage in the design cycle, rather while designing then afterwards. The larger computational power of computers will make this possible. Consequently, there will be less physical prototypes and thus less surprises when building the physical vehicle for the first time. CAD systems will hopefully be more "intelligent". Digital manufacturing is the goal.

4.4   Assembly Of Standard Parts

There will be less generic modelling features. It will be possible to simply assemble given parts.

4.5   Solution To PDM Integration Problem

The PDM integration problem will be solved either with STEP or other standards going beyond STEP. The software developers will all be forced in this direction. The global network of suppliers and car manufacturers will be sufficient to exchange the data they need, but this process will not be finished in 2007, it will take another five to ten years.